D0638366

Access Rules

Access Rules

FREEING DATA FROM BIG TECH FOR A BETTER FUTURE

Viktor Mayer-Schönberger
and Thomas Ramge

UNIVERSITY OF CALIFORNIA PRESS

University of California Press
Oakland, California

Library of Congress Cataloging-in-Publication Data

Names: Mayer-Schönberger, Viktor, author. | Ramge, Thomas, 1971- author.
Title: Access rules : freeing data from big tech for a better future / Viktor Mayer-Schönberger and Thomas Ramge.
Description: Oakland, California : University of California Press, [2022] | Includes bibliographical references and index.
Identifiers: LCCN 2021038723 (print) | LCCN 2021038724 (ebook) | ISBN 9780520387737 (cloth) | ISBN 9780520387744 (ebook)
Subjects: LCSH: Big data—Access control—United States. | Big data—Social aspects—United States. | Information technology—Social aspects—United States. | Technological innovations—Social aspects—United States.
Classification: LCC QA76.9.B45 M39 2022 (print) | LCC QA76.9.B45 (ebook) | DDC 005.7—dc23
LC record available at https://lccn.loc.gov/2021038723
LC ebook record available at https://lccn.loc.gov/2021038724

Manufactured in the United States of America

31 30 29 28 27 26 25 24 23 22
10 9 8 7 6 5 4 3 2 1

"Nothing in life is to be feared, it is only to be understood.
Now is the time to understand more, so that we may fear less."

MARIE CURIE

Contents

1. The Power of Information *1*
2. Data Alchemy *18*
3. Schumpeter's Nightmare *37*
4. Data Capitalism *55*
5. Might and Machines *80*
6. Access Rules *105*
7. Open Data Reloaded *126*
8. The End of Data Colonialism *150*

Acknowledgments *169*
Further Readings and References *171*
Index *183*

1. *The Power of Information*

In the early 1730s, a young publisher in Philadelphia would hand over two bundles to the city's postal couriers on a regular basis: a large one of copies of his newspaper, the *Pennsylvania Gazette,* and, covertly, a smaller one of bank notes. The money wasn't to cover the postage. It was a straightforward bribe. The young printer had no choice in the matter. His name: Benjamin Franklin.

Fifteenth child of a soap and candle maker, he had acquired the *Gazette* from Samuel Keimer, a flamboyant but heavily indebted publisher who fled his creditors for the Caribbean after a brief sojourn in jail. Ambitious and enterprising, the young Franklin was a brilliant essayist with a keen nose for the stories that caught the interest of a growing class of readers in prerevolutionary America. Under his leadership, his paper soon came to be seen as witty, entertaining, and politically smart while eschewing an overtly rebellious tone against the British colonial authority. But at the same time the paper was imbued with Franklin's conviction that a free press would pave the way to an American democracy.

The paper's quality, however, did little to help the talented publisher increase its circulation or influence. Unlike Franklin, no one would mistake the editor of Philadelphia's largest newspaper, the

American Weekly Mercury, for a brilliant essayist. But he did hold down a secondary job working for the British Crown. This was Andrew Bradford, Philadelphia's postmaster.

At the time, a colonial postmaster could decide at his own discretion which newspapers were sent free of charge through the mail—and which not. In a nutshell: A self-interested supernumerary of the King of England controlled the flow of information and its commercial use. The *Mercury* was sent by mail, and the *Gazette* wasn't. It was that simple. Benjamin Franklin was forced to play along. He managed to keep the *Gazette* halfway afloat by bribing the postal couriers whenever he could. But, in 1736, the situation changed.

The British colonial postmaster general had become increasingly dissatisfied with Bradford's services, especially with the meager profits he was generating. Consequently, he appointed Franklin, clearly a far more capable publisher, to be regional postmaster. Franklin immediately put an end to unfair competition in newspaper distribution. All newspapers in colonial Pennsylvania would now be distributed on equal terms. From then on, the *Pennsylvania Gazette's* circulation grew. Experiencing how the most important information channel in colonial New England could be abused shaped Franklin's outlook for the rest of his political life. And the postal system remained an issue dear to him. In 1757, Franklin was appointed co-deputy postmaster of the British Crown for all the American colonies. He was dismissed from this office shortly before the American Revolution for being too close to the rebels. During the Second Continental Congress, Benjamin Franklin advanced the cause of establishing an independent United States Postal Service. In 1775, he was named the first US postmaster general. The Constitution explicitly mentioned an independent postal service and made it a federal agency. The Postal Service Act, too, bore Franklin's handiwork. It required the postal service to deliver all newspapers—by far the most important source of information for citizens at the time—inexpensively and on equal terms throughout the country. The US Post became an irrevocable element of the founding myth of the United States.

What seems obvious today, after more than two centuries of experience with democracy, was already clear to the Founding Fathers: Access to information is the foundation of democratic decision-making. The principle of a free press means that journalists can report news and offer analysis and commentary and ensure that what they write reaches people.

Fast forward to spring 2020. As the first pandemic lockdowns were lifted, major decisions confronted politicians, society, and individuals: What can we now do again, and where? How can further waves of the pandemic be brought under control more quickly, more effectively, and, above all, in a more targeted way? This required information, and not just about the virus, but also about how it spreads and how people behave. Telecommunications and navigation system providers granted public health authorities access to regional mobility data, but that was neither granular enough nor did it help individuals to know whether and when they had been exposed to the virus.

What was needed was a way to trace individual infections. A number of Asian nations had weathered the first wave well; by manually tracking infections, they were able to quarantine exposed individuals effectively. Governments wanted to build on this experience but use a technological solution—smartphone apps. These apps were primarily designed to inform users when they'd had contact with an infected person. However, a number of countries also wanted tracing apps to provide information in anonymized form about how infections were spreading within a given locality. The hope was that, armed with such information, decision makers might avoid another complete lockdown and instead adopt measures more limited in time and scope.

With this goal in mind, public health authorities negotiated with Google and Apple. These two mega-businesses dominate the smartphone market. Their help and support are vital for tracing apps to work. Otherwise, such apps can't accurately measure the distance between users and won't be available for download in the app store. To the surprise of government officials, the Silicon Valley duopoly refused to help. Instead, the pair took on a role usually played by privacy activists—as evangelists of minimal data use. Google and Apple executives pushed for maximum privacy protection and, what's more, effectively answered a question normally left to democratically elected public officials who (one hopes!) are well-briefed by their scientific advisers: How best can we protect people's lives in a pandemic with all the possibilities on offer from digital technologies?

In May 2020, it became clear that Google and Apple would not provide authorities with the technical capabilities to get access to the basic public health information they needed to make vital decisions. Dominant digital platforms exercised a *de facto* veto over the democratic flow of information. Put simply, no longer could Boris Johnson, Emmanuel Macron, or Angela Merkel, or the governments of Japan, Australia, or South Africa, decide what information was available to whom and in what way to combat the pandemic. Instead, the decision was being made by Tim Cook and Sundar Pichai. Their decision crippled tracing apps in Europe, but also in some states in the United States, as well as countries like

Singapore and Australia. Apple and Google's actions may have protected individual privacy but left public health authorities without valuable data on how the virus spreads and where.

This book is about the power of information—and how power may shift as we change the rules on access to information. The frustrations of Benjamin Franklin remind us that unequal distribution of information, control over its flow, and resulting power imbalances are not just a phenomenon of the digital age. But the behavior of Google and Apple over tracing apps graphically illustrates how control over information in a data-driven world is shifting in favor of those who generate, store, and analyze information flows on their digital platforms. Since the days of Benjamin Franklin, information power has been turned upside down. Today, data colonialists in Silicon Valley, and to a lesser extent China, rule much of the world. In fact, these private corporations shape information access just as much within the United States, influencing economic transactions and democratic decision-making.

Machine-readable information, the wealth of data the Internet and smartphones have brought us, the rise of major digital platforms and of the superstar companies that create and control them, the digital collaboration tools we use and the traces of data we leave behind through them—all of these raise an old question in a new way: How do we legitimize—and limit—the power of knowledge?

Spoiler alert. In this book, we'll offer a clear, concise, and, we hope, compelling answer. We have to prize open access to information if we are to counter power asymmetries and eliminate undue information-based digital domination. We need much more broadly based data access to advance scientific, social, and economic progress in the service of sustainable development. Access rules. Because concentration of information power is good for a few, but bad for innovation, cooperation, and for each and every one of us.

Through our call to broaden access to information, we contribute to discussions about how one can contribute to an equitable digital transformation of our economy and our society. In many public and private debates, the question of informational power is rarely discussed. And when it does come up, the discussion is often one-sided and defensive, as if the answer to informational power is to be found in people's continued ignorance rather than their empowerment. Most of the time, though, the question of informational power is not even raised. We find this surprising for three reasons: First, it shows a lack of understanding of the nature of power; second, it doesn't do justice to the role information technologies play; and third, because policy responses to technology-driven imbalances of informational power are rare.

According to Max Weber, power is "every [c]hance, within a social relationship, of enforcing one's own will even against resistance, whatever the basis for this [c]hance might be" (formatting omitted). This includes information. The entire theory of innovation since Joseph Schumpeter revolves around the question of how informational advantages and advances in knowledge can be converted into market power. Manuel Castells refers to the postindustrial age as "informationalism," because it's so strongly shaped by the role of information and the power derived from it. In 1999, US economists Carl Shapiro and Hal Varian issued a guide for digital companies in the twenty-first century, explaining how they could use platforms and network effects to seize economic power. The title was not only revelatory for us—it was also inspiring: *Information Rules*. Superstar companies in Silicon Valley and, increasingly, Asia have followed this guidance—it's no coincidence that Google hired Hal Varian as chief economist in 2007 and went on to establish information-based dominance. When information rules, we need new rules of access.

Thanks to technology's influence on the distribution of informational power between organizations and companies, individuals and customers, and societies and states, the digitization and datafication of the world has yielded a series of dialectical developments. Over recent decades, whenever digital innovators have vowed a major technological leap forward, they always equally promised informational empowerment for individuals or small organizations. The personal computer (PC) supposedly democratized computing power, offering electronic data processing to the masses when once it had been reserved for big business and governments. The Internet opened the door to the world's knowledge for anyone with access to a networked computer. Google's founding mission was to organize the world's information and make it "universally accessible and useful." And social media, strongly supported by mobile Internet and smartphones, finally seemed to be taking away the keys from the old gatekeepers of informational power. The Arab Spring looked like an optimistic foreshadowing of how the exchange of information could fuel democratic debate and bring down dictators.

Each of these promises has in some respects been fulfilled. And yet every increase in the flow of information has resulted in a brutal countervailing force. The digital revolution intensified information asymmetries in a way that technological pioneers, from Alan Turing to Vint Cerf to Tim Berners-Lee, with their ambition to improve the world through technology, couldn't have foreseen—and certainly didn't intend. Put simply, in each wave of social unrest and political reform since the invention of the PC, information technology has been used to change social and economic structures and expand individual empowerment. But half a century later, these power structures have not only failed to crumble they have solidified in favor of centralization. All that has changed is the names of those in power. It's no longer oil barons or bankers sitting on top of the economic power pyramid but Tim Cook and Satya Nadella, Jeff Bezos and Mark Zuckerberg, Larry Page and Sergey Brin, and in China Baidu's Robin Li and Tencent's Pony Ma.

Their power stems from their ability to collect and analyze digital information, to keep exclusive control over it or dole it out if and when it suits them. To paraphrase Max Weber, they can advance their interests even against the resistance of others thanks to the power of their data. Thinking back to young Benjamin Franklin and his struggle to get his newspaper into circulation against the arbitrary whims of his state-sanctioned competitor, it seems like a bad joke of history that today's Big Tech is competing against publishers for advertising income while publishing their content and thereby inhibiting profitable quality journalism. It's not against the law, of course. And yet it undermines the dissemination of and access to information at the heart of technological, economic, and social innovation.

The shift of economic and media power—and political power—in favor of data-rich platforms has been mapped. In the United States, the debate has been advanced by, among others, Shoshana Zuboff with her interpretation of "surveillance capitalism" (and the unreasonably less popular Julie Cohen's *Between Truth and Power*), Tim Wu with his work on net neutrality, Eli Pariser with his formulation of filter bubbles, and Roger McNamee's settling of accounts with Facebook and the computer industry in general in his book *Zucked*. Experts, such as the antitrust scholar Lina Khan argued for strong regulatory measures. Adrian Wooldridge, columnist for the *Economist,* summarized the growing unease with Big Tech's information dominance by suggesting the term "techlash," the angry reaction of consumers and regulators against Big Tech's power to control people and markets.

By 2021, Big Tech faced antitrust activities by regulators and policy makers across the globe. In the United States, forty-six of the fifty states had sued Facebook, accusing the social media behemoth of anticompetitive conduct. At the same time, Facebook also faced legal action from the Federal Trade Commission. Google, too, was sued by dozens of states for anticompetitive behavior, in at least three different lawsuits. The Trump administration had long and loudly criticized Big Tech for its alleged bias against the president. His successor, President Joe Biden, had not made Big Tech a campaign issue, but in office he swiftly nominated key critics of Big Tech to powerful positions in his administration: Tim Wu became Special Assistant to the President for Technology and Competition Policy and Lina Khan was appointed Federal Trade Commissioner.

In Europe, the Danish liberal politician Margrethe Vestager had pushed antitrust actions against Big Tech for years as Commissioner for Competition. Under her leadership, the Commission levied fines in excess of $8 billion on Google (although the European Court of Justice reduced many of these fines on appeal). In December 2019, she was rewarded for her active stance by being elevated to the role of Executive Vice President of the Commission, promptly promising more antitrust action to come. Meanwhile in China, in late 2020 the government began its crack down on Big Tech, initially focusing on Alibaba and Jack Ma, Alibaba's charismatic founder. First, the Chinese authorities prohibited the much-anticipated IPO of Ant Group, which runs China's dominating digital payment platform Alipay. Then in April 2021, Alibaba was fined $2.8 billion for "anticompetitive practices."

These regulatory actions seem impressive and focus on the right issue: rebalancing information power. But we doubt that they will have sufficient impact. Much of the analyses of the current gross imbalances in informational power has led policy makers and regulators to conclusions that are just as one-sided and reactive, as they often have been when critiquing digitization. The responses are still far too defensive in theory, and inefficient in practice. Perhaps the most visible example of such an approach has been Europe's General Data Protection Regulation (GDPR), intended at least in part to hobble Silicon Valley data giants and give Europe's citizens sovereignty over their personal information, and lauded by some Big Tech critics in the United States as an encouraging example of how to curb information power.

But despite its noble intentions, GDPR has in fact helped digital superstar companies enlarge their informational power and expand their centrally planned digital economies. In the age of data capitalism, allowing data to be processed solely with an individual's consent has unintended and unfortunate consequences. Because, in practice, individuals volunteer their personal information to digital platforms. Utilizing a multitude of these digital services overburdens us with the task of managing our data protection settings judiciously. So, instead of rising up against data-hungry Big Tech, we carry on using their online services because they're convenient and useful. The price we pay is individual and economic dependence in spite of—and in part even *because* of—privacy laws, as they push much of the responsibility for when and how to use personal information onto the individual.

Information asymmetries continue to grow in favor of digital superstars, despite numerous attempts around the globe at regulation. The result is sobering: Global information giants manage just fine. Technologically and legally, they've mastered the complex regulatory environment, while companies lagging behind digitally are driven to despair by the bureaucratic burden. As a consequence, they use data even less. Viewed objectively, the GDPR that many European data protection advocates are so proud of (and that numerous privacy activists elsewhere want to mimic) is facilitating the power of digital giants. And behind closed doors in Silicon Valley and in China, you'll often hear, "Isn't it weird Europeans don't even notice how they're shooting themselves in the foot?" It isn't different in other markets with stringent data protection laws either: consumers provide their valuable personal data to (mostly foreign) Big Tech, while domestic digital startups are hampered by bureaucratic processes.

In this book we are critical of the role current data protection regulation plays vis-a-vis Big Tech. But our central message is far bigger: Ultimately our societies have to develop a pro-active rather than defensive strategy when it comes to the concentration of informational power. It is up to regulators and legislators, and ultimately voters, to compel information giants, wherever they are located, to share their treasure troves of data with others. We have to open access to information to all: citizens and scientists, startups and established companies, as well as the public sector and nongovernmental organizations.

In our era, with a tech cold war raging between the United States and China, and with hybrid cyberwarfare emanating from

Russia, worldwide open access to data may seem like an idealistic utopia with no realistic chance of coming to pass. But if we imagine and then enact an alternative regulatory framework based on broad access to data, a framework that enables innovation, fosters competition, and delivers a data dividend to society, others not only in the West, but among emerging economies in Asia and the global South may join in.

In the second and third chapters of this book, we'll describe why data has become the most important raw material for innovation, and how dominant data oligopolists, contrary to their reputation as digital pioneers, are in fact slowing down innovation—for the benefit of their shareholders and at the expense of customers, the economy, and society. But there is an opportunity for digital liberation. Powerful informational tools in the hands of a few could instead become engines of empowerment for everyone conversant with information technology.

Then, in the spirit of economist Joseph Schumpeter, both national economies and societies could reinvent themselves as data innovators. And instead of losing top data scientists and machine learning specialists to Silicon Valley monopolists, such talent could help nations near and far to establish and regain business dynamism by transforming data into insights.

Admittedly, such reinvention is no easy undertaking, neither mentally nor technologically. It is possible though as long as there is sufficient political will. Legally it would be easier to implement than one may think. Any company would be required to give everyone access to its data. It would be a precondition for doing business. Of course, data would be stripped of personal identifiers. And trade secrets and similarly protected data would be excluded. But research-related and anonymized health data for instance would be allowed to flow freely. The public sector would also be required to open its data stores. In the fifth chapter, we'll describe in greater detail why an open data space might result in short-term loss of market share and profit margins for some, but why the overall long-term economic and societal gain could be enormous. And the political desire to break up information monopolies and turn powerful informational tools into instruments of empowerment is growing.

When we promoted the idea of mandated data sharing in our 2018 book *Reinventing Capitalism in the Age of Big Data,* some people looked upon us as make-believe artists. But in Brussels, mandating access to data is now being discussed with increasing intensity. There are also encouraging signals from emerging economies that could lead to a global debate about access to data. Those responsible for digital policy in populous emerging economies such as India, Pakistan, and Nigeria—where the very size of their populations gives them a super-abundance of data—are increasingly coming to realize that asymmetrical informational power is reviving the postcolonial conflict between the global North and South. Their dependence on digital services from Big Tech is growing by the day. In the concluding chapter of this book, we describe how the people in these economies are providing huge volumes of valuable data, including through poorly paid clickwork. And how Big Tech is skimming off economic value. It's striking that for the first time in history, the former colonies of the South are in the same boat as the former colonial powers of Europe ruled over by Big Tech's "data colonialists." And perhaps even for the first time in postcolonial history, large emerging economies might be better able to reset the power relationship thanks to a clear stance and a stronger starting position compared with the West.

Meanwhile, the empire of data is preparing for a long and bitter defensive battle. A tell-tale sign is how Apple, Google, Facebook, and Amazon are expanding their armies of lobbyists in Washington, London, and Brussels at breakneck speed and with considerable urgency. The central issue is no longer data privacy. The superstar companies know they can easily handle this policy issue. Instead, their lobbying agenda is now focused on ensuring that they retain exclusive use of the data that users from all over the world have left on their servers. They don't say it out loud, but they have come to realize that data protection as we now understand it simply helps them achieve that goal.

Broad access to data is not a far-fetched dream. It's an achievable vision. For the vision to become reality, all it takes is courage to give it flesh. But to achieve this, we will have to bid farewell to the principle of minimizing data use. We have to understand that a data *use* regulation is as necessary for prosperity and democracy as data protection regulation is for protecting individual rights. They're two sides of the same coin.

Like knowledge, data has an amazing quality. Economically speaking, it's a "non-rivalrous good." That makes it more suitable to be a public good than any other important economic resource so far. "Data for everyone" won't end in failure simply because everyone starts grazing their cows and all the grass gets eaten. As a digital publicly accessible good, data doesn't disappear as more and more people use it. That's because data turns into value only by being used. And its value increases with every additional use. It's simply foolish to let a few data-rich corporations limit the value and insights society can gain from data. We don't even have to expropriate corporations, since in a legal sense data cannot be "owned"; we only need to rewrite data access laws.

Studies show that more than 80 percent of all data being collected isn't even used once. The main reason—those who could create value by using it don't have access. Data monopolies are theft of progress. Data usage is service to the common good.

This is a vision that has been eloquently articulated by Italian innovation economist and digital policy expert Francesca Bria. Bria argues that regaining access to data by individuals, groups, and society can indeed realize a much broader vision of not just economic but also societal reinvention. She urges societies to "treat data as a common good." Her vision to counter what she calls Big Tech and Big State is Big Democracy—a "democratizing of data, citizen participation and technology at the service of society."

By advocating for the democratization of data, Bria isn't simply offering a powerful vision. She also has been successful in turning that vision into reality. After her academic work on innovation and a stint at NESTA, the British innovation agency, Bria ran digital technology and innovation for the city of Barcelona. There, she persuaded digital platform providers, such as telecom incumbent Telefonica, to open access to its data troves to individuals and local businesses, gifting Barcelona with a digital dividend that facilitated economic development and societal empowerment.

Visionaries with the ability to shape reality like Francesca Bria lead the way. The time is ripe to take back power from the information giants, who only grew to such dominance through the use of "our" data. Over the past decade, the pressure on Big Tech has continued to build. But perhaps it takes a disruptive event like a deadly virus to make us rethink access to information. The pandemic has opened our eyes in two ways. First, we discovered that our systems are significantly less resilient than we thought. Second, when we look at the development of vaccines, drugs, and public health measures, we realize we have only been able to control the virus through the unusually open exchange of information.

In the face of a deadly threat, what we previously thought impossible became possible. Scientists around the world shared their knowledge more rapidly and more generously than ever before in the history of science. Competing drug makers formed new alliances in their search for tests, therapies, and vaccines, because they recognized that only through information sharing and cooperation would they be able to work at the speed that was required—ultimately for the benefit of everyone's health (and, of course, their own commercial returns).

Even Big Tech companies pitched in. With its gigantic logistics machine, Amazon did a good job providing basic necessities to many people who couldn't leave their homes during lockdown. Apple and Google published anonymized movement data calibrated geographically so that public health officials could better assess the effect of partial lockdowns. Microsoft initiated an open data initiative with the explicit goal of supporting innovations that benefit everybody.

Two years after doctors in Wuhan first diagnosed the symptoms of an unknown lung infection, it's now more evident than at any time since World War II: we will only be able to make our world more resilient against major crises through better information-sharing. The basis for this is the free flow of data. The age of information monopolies is over. This book describes the path toward a new era, in which people all over the world will always have the digital tools to access the data and knowledge required for finding solutions to the great social, economic, environmental, and health challenges that confront us. We will no longer ration data artificially through overly complex regulations and we will recognize that data monopolies are undermining competition at the expense of customers and the state. Data-rich nations around the world will open access to reap their data dividend.

And we will all understand that the benefit of data emerges from its use. The more often we use data, and the more diverse our purposes in doing so, the greater the economic and social value we'll reap. The world will first perceive and then measure precisely how data for all is good for everybody, when access rules.

2. *Data Alchemy*

March 2004. A motley crew of students, tinkerers, and engineers gather in a restricted military area in the Mojave Desert. They proudly show off the futuristic autonomous vehicles they've built. Some of them could be mistaken for clones of *Star Wars* spaceships on six wheels. Others look like small tanks whose guns have been replaced by bulky laser sensors. There's also a self-driving motorcycle, packed with electronics and balancing weights. However, many design teams have come with conventional pickup trucks from Ford, Dodge, or Toyota. Cameras are mounted on roofs and on front sections. In the passenger compartment, bulky computers compete with people for space.

The sun is shining, as it almost always does at the border between California and Nevada. Most of the participants are dressed in nerd uniform: shorts and T-shirts or polos with the logos of tech firms or universities. But there's also a conspicuously large number of military personnel bustling about in the designers' encampment. The event's host is the Defense Advanced Research Projects Agency (DARPA), the main research arm of the US Department of Defense. Even if there's a collegial atmosphere among the teams, they've come to the Mojave Desert to compete.

The organizers have marked out a 150-mile course for the "DARPA Grand Challenge." If a self-driving vehicle completes the circuit without a helping human hand, its inventors will win a million dollars.

At the start of the race, spectators cheered on their autonomous vehicles like marathon runners. But the technology turned out to be closer to "couch to 5K" in fitness terms. The self-driving motorcycle toppled over after just two yards. Many vehicles could still be seen with the naked eye when they came to a stop. A converted Humvee from Carnegie Mellon University went farthest. The "Sandstorm" vehicle reached beyond the seven-mile mark before it got stuck on a rock while negotiating a bend. At that moment, the dream of autonomous driving seemed impossibly far away.

Today, almost two decades later, the technical hurdles for driverless cars have largely been overcome. In February 2021, the California Department of Motor Vehicles released data on test drives conducted by major technology companies, ride-hailing providers, autonomous driving start-ups, and traditional automakers during the previous year. The so-called "disengagement rate" is one indicator for the technological progress of self-driving cars. It tells you how often on average a safety driver had to switch off the autopilot and take control of the vehicle. The self-driving cars of Google's Waymo subsidiary drove over six hundred thousand miles on California's public roads in the reporting period (December 2019 to November 2020, which includes the pandemic lockdowns during which some self-driving tests were suspended). On average, safety drivers had to grab control of the steering wheel just once every thirty thousand miles. It was a huge leap forward. Only five years earlier, Waymo vehicles had gone less than two thousand miles without human intervention.

No other company did better, neither General Motors' Cruise nor China's AutoX or ambitious start-ups like Zoox or Pony.ai. Lyft and Apple vehicles only drove a few hundred miles on average before a human needed to take over. And Tesla, who had reported abysmal results in 2019, did not even provide figures for 2020.

More revealing, however, was the fact that seasoned car manufacturers, including well-known German and Japanese brands, were at the bottom of the heap, achieving less than a hundred miles before disengagement (and driving only a few hundred miles throughout the year in total). Given such dismal results, these automakers will likely have to license the Waymo system. Or perhaps they'll opt for another competitor that until recently had only caught the eye of a few automotive experts but one well-known to Google since it had replicated its technology and business model once before: Baidu, the Chinese search engine giant. Since at least 2017, Baidu has been investing heavily in autonomous driving with a program called "Apollo." Apollo vehicles have been tried out on Chinese roads over several million miles, but Baidu was also testing a small fleet in California. In 2019, its vehicles there traveled on average more than fifteen thousand miles between human interventions, second then only to Waymo, although experts questioned the validity of these figures. They suspected Baidu achieved this feat by driving mostly on California's highways instead of on city streets, where driving is more difficult, and human backup drivers are generally forced to intervene more often (Baidu failed to report figures in 2020).

What's undisputed, however, is that Waymo's and Apollo's systems also improve at a faster rate than their more traditional rivals. There's a crucial reason for this: Google and Baidu have succeeded in creating a technical environment in which their vehicles generate and utilize more valuable training data than those of their competitors. With every mile driven, with every intervention by the safety driver, with every tricky situation mastered without intervention, vehicles learn. They are then able to drive farther and generate new training data more easily. Google cars drove about 350,000 miles in California in 2017, more than a million the following year, and about 1.5 million miles the year after. Their approach is paying off. They can invest in larger fleets and at the same time set up virtual worlds based on real data collected from their test fleets.

In these simulated worlds, digital twins drive billions of miles a year—in Google's case aided by technology from DeepMind, its machine learning subsidiary. The data from California represents only a small part of Google's entire autonomous driving efforts. Self-driving Google vehicles are cruising public roads in twenty-five cities in the United States—mostly in sunny Phoenix, Arizona. Some of the vehicles are already functioning as real-life robo-taxis without safety drivers. At the same time, Baidu began offering an experimental fully fledged robo-taxi service, called Apollo Go, in Beijing in September 2020. When vehicles are operating without human backup, we've crossed a threshold that medieval alchemists could only have dreamed of. Self-driving cars are becoming self-learning cars. On their own, the vehicles generate the raw material with which they improve themselves through machine-readable information.

Self-Learning Machines

Such data alchemy is the apex in the field of machine learning, the most important area of AI development in recent years. Despite all the hype about AI, the impact of machine learning remains *under*estimated. If, after an initial training phase, systems that learn from data can themselves generate data for refining their algorithms and improving applications, they can part-automate the process of innovation. In the next chapter we'll take a detailed look at what this means for market concentration. For now, it's important to understand how machine learning increases information asymmetries and affords a significant advantage to the suppliers and operators of systems that learn from data. For users, these asymmetries lead to unwelcome dependencies. For competitors, it's all but impossible to catch up with leading providers.

The evolution of self-driving cars from Google and Baidu is a particularly striking example of data alchemy. But we can find other examples in almost all applications where digital engines harness the power of machine learning. With every word we type into a search engine, the system gets to know us a little better. With every click on a result, we provide feedback to the search engine: This bit of information is relevant to us, that bit isn't. If we don't click at all, that's also a valuable signal, of course. The more often we and countless others use the system for searching, the better it can calibrate its algorithms and the more likely it is to provide us (and everyone else) with relevant results. The recommendation algorithms of major online retailers adhere to the same principle. The algorithms are in effect product searches that Amazon or Ocado, Flipkart or Alibaba are happy to carry out for us. The more customers follow the recommendations (or the more who pointedly reject them), the better the retailer's computers can use pattern recognition and machine learning to optimize its product range, calculate prices, and plan marketing campaigns.

Cancer diagnosis software based on machine learning improves with every diagnosis carried out as the result is fed back into the system. Credit card providers' fraud detection systems learn from each suspicious transaction they block or authorize. A customer who's been defrauded can be counted on to provide angry feedback, as can an honest customer left standing at the register like a fraudster because her credit card was declined. The more data a bank's credit scoring system has about payment defaults, the more accurately it can predict whether a particular loan applicant is likely to repay.

Speech recognition software understands spoken words more accurately the more often people address it and correct misunderstandings. The more interactions recorded between machines in a smart factory and the more learning experience its central control software accumulates, the more efficient it becomes. The more contracts a legal bot checks, the more often we can save ourselves a costly consultation with a lawyer. Why? Because modern computer systems and human beings both learn in a similar way. They collect information, evaluate it, and then draw the right conclusions using data-based predictions. With built-in feedback loops, they automate information gathering and learning effects. The principle of data alchemy in systems that learn from data is only the latest feat in a much larger and deeper shift of economic power caused by information technology.

Data Is Not Like Oil

The rise of digital superstar companies with their agile organizational charts and disruptive business models has been much analyzed and debated. CEOs of traditional companies and the consultants advising them have racked their brains about how to adopt the superstar model. And when they make an often-feeble attempt to do so, they call it "digital transformation." Terms like "Big

Data," "advanced analytics," and "AI" feature prominently in their discussions. And yet the fundamental differences in the use of digital data between superstar companies with data-centric products and business models and the large traditional companies in their approach to digital transformation are often—and oddly—given scant critical consideration.

In traditional sectors such as machine building and automotive, financial services and telecommunications, restaurants and retail chains, consumer goods and logistics, data was and is generally understood and used as a resource for optimizing processes and products in a predetermined way, or as a lubricant for the marketing machine—that is, to sell more products to customers. That makes sense, of course. Wherever a company can increase efficiency, reduce costs, and increase sales with data, it should do so. But this functional approach to data is fundamentally different from the strategy of the superstar companies that emerged and succeeded in the Internet age.

Today's most valuable companies don't see data as an excessive burden, as an expense to be addressed through cost-cutting, or as a quantifiable lever for increasing sales. For them, data is a key investment in their own future. It's no coincidence that the skewed metaphor of "data as the new oil" was popular with managers in old industries. The image fits the concept of conventional value creation: Existing raw material only needs to be extracted; then it has inherent value and earns money when you sell it. The digital champions, on the other hand, never saw themselves as the ExxonMobils or BPs of the data age. They knew that data doesn't slumber deep inside a mountain waiting to be extracted like gold ore or pumped out of the ground like oil. They knew that data doesn't burn when used, resulting in high externalities, but rather its value increases precisely through diverse, repeated, and combined use. And it was clear to them early on that the way to add the greatest economic value was not selling data but using it to establish domination based on exclusive access, which in turn drives their highly profitable business models.

Platform Hermeneutics

Today's data-rich companies have spent considerable effort in establishing a technical infrastructure that directs the flow of information to their servers and ensures exclusive use of it. Achieving this strategic goal is a historic business event. Wherever we take a closer look at the mechanisms underlying the digital economy, we find this pattern at work in creating information asymmetries. The mechanism is particularly evident in the powerful digital platforms that have elbowed their way in between buyers and sellers in so many industries.

All transactional information generated by smartphone app stores is in the hands of Apple and Google. The outcome is less alchemy than hermeneutics. A single detail demonstrates value only when viewed within the bigger picture, but the latter can only be seen by someone who knows all the details. App store providers know their individual customers, their preferences, and their habits. By pooling this information, they put together a comprehensive but granular image of the overall market and can then align their business model accordingly, whether it's for creating and marketing apps, for selling content such as music and videos, or for personalized advertising.

Amazon Marketplace not only collects sizable fees from "independent" retailers who offer their goods through the platform; the information derived from these countless transactions also enables Amazon, with its commanding view of the market, to sell the most profitable products itself, or give its even more profitable own-label brands pride of place in their marketplaces. On travel platforms such as Booking.com, delivery portals such as Uber Eats or DoorDash, or with (seemingly) neutral price comparison services such as PriceGrabber, the business model, too, is based on unequal information power. With access to a treasure trove of data, the digital matchmakers know which customer is looking for what special offer. Often, for the sake of convenience, customers satisfy a platform's thirst for information, while vendors, feeling powerless, permit a platform's data gathering behavior in the absence of viable alternatives. Viewed objectively, however, the unpopular intermediaries simply invested enough money at the right time in a digital marketplace that is convenient for users and provides added informational value. The marketplace collects data on supply and demand, combines it in return for a (often significant) fee, learns from each transaction it brokers, and therefore, with its increased knowledge, is able to match supply and demand ever more precisely and thus earn even more in fees.

Nobody stopped the hotel industry, retailers and supermarket chains, restaurants or taxi trade from building their own platforms to help customers find and book their services and products. Nobody prevented the pharmaceutical industry from offering a low-cost individual gene sequencing service like 23andMe, and with each new customer—the Californian start-up can count more than ten million of them—add additional troves of valuable data to its database to license to others.

Major suppliers of medical imaging technology had the resources to develop AI systems for cancer detection in collaboration with recent graduates in data science. The data would presumably have been even more easily accessible to them than to the machine learning experts at IBM. But General Electric and its traditional competitors surrendered the field to Big Blue and other tech firms without much of a fight.

Telecommunications companies could have introduced messenger services that today are evolving into social networks, or, like Google, could have trained voice recognition software by offering a free telephone directory. But instead, the telecom companies wanted to continue selling old-fashioned SMS, and they shut down their own directory services altogether.

Hollywood studios were sitting on mountains of content when Netflix was still laboriously mailing out DVDs. But Netflix built a digital platform. The company analyzed video usage data to discern how various storylines played with viewers. It turned itself into a vertically integrated producer and provider of addictive TV shows. Initially late to the party, Disney is now betting on its own platform to establish a direct relationship with customers. Banks in the West and Far East have left the digital payment market to services such as PayPal, Stripe, Wise, Alipay, or WeChat. Not banks, but fintech (financial technology) companies like Robinhood and Chime in the United States, Klarna and N26 in Europe, and Ant Financial in Asia, entered the market, while the banks themselves remained monoliths, possessing an extremely large amount of data, but rarely making use of it and, instead, often fighting against their own grotesquely outdated IT infrastructures.

And who would have had the money (and auto fleets) to re-measure the world's road networks with vehicle-mounted cameras, thus collecting data for advances in autonomous driving (that we described earlier as a self-learning system), if not Toyota, Hyundai, or Mercedes, you'd have thought? But while automakers were busying themselves on making the latest seven-speed automatic transmission a little better than the one before, Google was dispatching its Street View cars. Why? Because data is not the new oil. It isn't stored in tanks or, as the existing business model sees it, "consumed." Data is something that unfolds its value by being used repeatedly. This enables information, innovation, and economic power on a scale the oil barons of the nineteenth century could only have dreamed of.

All this isn't necessarily the fault of the former giants of industrial and financial capitalism. Switching to information-centric business models has never been as easy as highly paid change consultants claimed. Harvard economist Clayton Christensen highlighted the dilemma of many successful companies constrained by path dependency: as the pressure to reinvent themselves increases, they also have to serve increased customer needs and earn money. It's well known that there's no way out of dilemmas without paying high cost.

It's always the wrong time for incumbents in automobile, financial services, retail, or media to radically change business models, processes, and product portfolios. As Christensen predicted in his 1997 bestseller *The Innovator's Dilemma*, this risk can only be taken by disruptors backed by venture capital. By and large, Christensen has been proven right. That's precisely why, if anything, it's even more ironic that the major disruptors of the last twenty years, despite constantly showing up at public conferences and ostentatiously flaunting their (alleged) transparency, have been able to hide the secret of their success.

Superstar Quants and the Half-Truth

How does Google work internally? In 2014, the then CEO Eric Schmidt and Senior Vice President for Products Jonathan Rosenberg offered an answer in their book *How Google Works*. Over more than 300 pages, they offered entertaining anecdotes on company culture, agile team-working by brilliant minds, creative collaboration, and hunger for the next leap in innovation so typical of Silicon Valley. Schmidt and Rosenberg painted a vivid picture of how the company attracts talent from all over the world. They praised evidence-based decision-making and, associated with it, a deeply rooted mentality favoring experiments and tests. The book's publication was accompanied by YouTube videos of inspiring presentations by the authors (even freely downloadable PowerPoint slides were made available, which organizational consultants and New Work advocates could easily incorporate into their own talks).

How Google Works was an ingenious marketing coup with a clear message: Google had become one of the most powerful companies in the world because two young geniuses figured out how to build a better search algorithm at just the right time; and then, located on a congenial campus and organized in a network structure, many bright young minds were allowed to collaborate much more creatively than the army of employees in the top-down organizations of the old, pre-digital world. Good storytelling does not mean telling fairy tales. Presumably, most of the anecdotes in the book are true, and of course the workplace culture of Silicon Valley has been an important element of success in the rise of Apple, Google, Facebook, and other tech giants. But good storytelling doesn't tell the whole story. There's almost no mention in the book of information asymmetries or the exclusive use of troves of valuable data.

This isn't just Google's strategy; basically, all superstar companies have followed the script. They shared knowledge and technology in areas that weren't decisive in the battle for digital dominance. Apple knows exactly who installs which apps on their iPhone and when but doesn't share this knowledge with app developers. The same is true for media content on Apple News. And without access to data, content suppliers are reduced to commodity service providers. They're at Apple's mercy.

Spotify's method is very similar. The Swedish platform knows what music its roughly 350 million users are listening to and when, but it guards this data as carefully as Coca Cola guards its original recipe. And it uses this information asymmetry to maintain the upper hand in all negotiations with suppliers. Booking.com lets us know many things about the accommodation it offers. But the company also knows details of the capacity- and time-driven pricing algorithms used by hotels on its platform, because it observes them in use every single day. Nobody can get hold of that data because it's the root of Booking's information power.

The same is true for WeChat, Tencent's social media app that's hugely popular in China. There, data from one aspect of the platform is extensively used to improve the service in other parts, from messenger to social media and payment. At the same time, WeChat carefully guards the data it collects, fully aware of the tremendous information advantage the data provides. In this sense, WeChat, too, is following the familiar script—much like many other leading Big Tech companies in China.

In retrospect, it seems innovators relied on a rhetorical smoke screen on their paths to superstar status. They were able to hide the real source of their disruptive success until it was too late for incumbents to react. To explain the development and consolidation of their informational power, they've developed a particularly misleading narrative. The story goes something like this:

Digitization granted the world an incredible amount of data. Infographics show the (allegedly) exponential growth of data, measured in zettabyte. A zettabyte is a number with twenty-one zeros. Of course, no one can picture this, but that does help the misleading narrative: It makes it obvious that traditional companies are completely overwhelmed by this flood of data. They don't have the supercomputers to handle zettabytes. And even worse, they lack the smart guys—the AI specialists, data scientists, and quants—who are able to detect and keep the hidden insights of Big Data thanks to their secret knowledge of algorithms. Computing power and the required human intelligence, under this narrative, are unevenly distributed in the economy and concentrated among digital champions. Their dominance in information technology is the consequence of this dual inequality.

The story sounds convincing, but it's a hoax. It's based on an understanding of technology that's flawed in at least three ways.

First, the computing power needed today to establish informational power is available not only to a few large companies and governments of rich countries, but to almost everyone who can pay. Thanks to Moore's Law, processing capability has increased dramatically over the past sixty years, as has the capacity to store and transmit information. That doesn't just mean we can carry out more calculations than before—and faster. Moore's Law has above all reduced cost. This trend, which has persisted for decades, has been reinforced by Bezos' Law. Alluding to the success and aggressive pricing strategy of Amazon Web Services' (AWS) cloud division, it describes how and why cloud computing is democratizing access to information processing at high speed and in large volumes.

The costs for cloud computing on Amazon Web Services are halved roughly every eighteen months. All things considered, and put in perspective, this means: In the 1960s there were only a few computers; hardware was highly concentrated. That concentration is now a thing of the past. Not because everyone has a smartphone in their pocket, but because smartphones, tablets, laptops, and, above all, the computer networks of companies of all sizes permit access to the cloud, and thus to almost unlimited computing power at minimal cost.

It's something not everyone has grasped fully yet. False perceptions of the role and importance of computing power can be observed, for instance, in discussions of the development of GAIA-X, a nascent European infrastructure for cloud computing. Funded through European tax revenues, GAIA-X initially was imagined as a direct competitor to cloud providers from the Silicon Valley (and, in the case of AWS and Microsoft, the Seattle/Puget Sound area), but as public subsidies were cut back, GAIA-X proponents invited US cloud providers to join, realizing perhaps that constructing a European cloud infrastructure would be anachronistic. The problem isn't that there are too few cloud providers in the world, but the opposite: there are too many.

It's a lesson China is learning the hard way as well. With enormous private and public sector investments (more than $13 billion in US dollars just in 2019), the country established cloud service providers that are still hoping for spectacular profitability as they expand aggressively internationally, especially in Asia and the Global South.

We know from economics that infrastructure services can either be provided by a monopoly, and therefore turn out expensive (unless regulated by the state). Or multiple infrastructure providers can be in fierce competition with each other. Then, the service turns into a commodity and profit margins are small. That's the case with cloud computing almost everywhere. Many cloud providers operate the equivalent of "dumb pipes." They may process data, but they have no access to the value that is generated through data use on their servers. Using public funds to build dumb pipes is simply wrong-headed. And it won't get us closer to "digital sovereignty" either (we come back to this issue later).

The second widespread misconception about why digital superstars achieved superiority is the ill-understood role of algorithms. The story depicts these as the true intellectual gems of the digital superstars and their amazing engineers, intellectual property worth guarding better than the Kremlin guards Lenin's corpse. Only the competing pantheons of Google/Apple/Facebook /Amazon and Baidu/Alibaba/Tencent, the story goes, have the mathematical wizardry to reach data wisdom. But fact is: That's pure nonsense. Many of the widely used algorithms for data analysis have been developed by academic researchers. Often, they're published and freely accessible in open-source algorithm libraries. For example, the highly in-demand data analysis software known simply as R is not only an open-source project but also available for download free of charge. A consortium of developers, primarily at universities, oversee and contribute to the project. The same is true for many common methods of machine learning. Some of the most important ML algorithms are published and have been known for years. Corresponding tools and applications are available on open platforms such as GitHub (now owned by Microsoft). In fact, when it comes to algorithms, superstar companies are astonishingly forthcoming. If they do develop algorithms themselves, they often keep them proprietary only for a limited period of time. And in some places, Silicon Valley is even taking a step toward openness. Founders like Elon Musk are financing platforms like OpenAI, where anyone can download well-regarded machine learning tools for free and put them to use.

The third and perhaps most awe-inspiring element in the popular narrative of how digital disruption caused a shift in informational power is human ingenuity, the suggested true reason for success. In popular science fiction, there's a long tradition of the genius who built an astounding invention that changed everything. In the digital age, this stock figure arrived in the roles of the ultra-nerdy high-IQ couple Dr. Amy Fowler and Dr. Sheldon Cooper from the cult classic *Big Bang Theory* TV. In the narrative of the superstar companies, their campuses are teeming with little Amys and Sheldons who turn data into gold with Einstein-like brilliance.

But the truth is, while data alchemy as described earlier may exist, there is no secret cabal of data alchemists. The data scientists and quants of major digital companies are using the same mathematical and statistical ingredients as everyone else. The dominance of Silicon Valley giants isn't due to the extraordinary human intelligence employed by these companies. When it comes to science and knowledge of methods and models, others, including start-ups, in the United States, Europe, and Asia can keep up. Just like they can when it comes to well-trained data scientists.

So, what does this all mean? Companies and economies that have been digitally left in the dust don't lack computing power for cloud applications. There's no shortage of tools for data analysis or algorithms for machine learning. And talent is also more widely distributed around the world than both digital superstars and grumpy latecomers to the party would have us believe. What's lacking for a major digital leap forward is access to data, the raw material of the digital revolution.

Google, Facebook, Amazon, Apple, Microsoft. Netflix, PayPal, Spotify, Uber, Booking.com. Baidu, Alibaba, Tencent, ByteDance, SenseTime, and Yitu Technology. The digital giants have created mechanisms, opportunities, and applications in their fields to collect massive amounts of data that they alone control. There is indeed a parallel here between data and oil. Whoever sits on the oil field doesn't have to give anyone else access. The data giants share their raw material with others only in exceptional cases. This is understandable from a strategic perspective, because access translates into competitive advantage; it's the basis for how their business is valued; and it leaves their competitors bewitched, bothered, and bewildered.

The digital innovators, who turned into Goliaths in their own right, have every reason to keep alive the story of their ascent as an astonishing tale of human ingenuity and the algorithms it generated. The narrative of achievement provides the moral legitimization of their power. It insulates them from any critique of information asymmetries; it protects against pointed questions. After all, who would want government regulation to rob the genius entrepreneurs of their success? The well-paid engineers at digital superstar companies believe they're working to advance information technology. And often they are sincere in their desire to permit others to access the fruits of their labors. At scientific conferences, they appear as intelligent and likeable, bathing in their status as superstars. Their management revels in sending their "Fowlers" and "Sheldons" to academic events because their attitude and their shared brilliance contribute so well to the overarching narrative. All the while, the people manning the command of digital superstars hype their technology as well as the skill of their employees, but they remain silent about data access.

This isn't new. Just take Thomas Edison, a fan of direct current in electricity, who tried to discredit alternating current technology by using tricks, stories, and schemes of all kinds. He even argued for using alternating current to execute people publicly. Again and again, those in power have tried to divert attention from the real source of their power by using intuitively plausible stories. Now with datafication, history is repeating itself. This time what's at stake is more than economic power. It's informational power—it's about dominating the control room of modern society. That's exactly why the deeply flawed story of honestly earned informational power is so dangerous. It protects the powerful, in Schumpeter's words, from the very form of creative destruction by which today's information giants themselves came to power.

3. *Schumpeter's Nightmare*

As a young man, according to a story Joseph Alois Schumpeter liked to tell later in life, he had three major goals. He wanted to be the world's greatest economist, Austria's top horseman and Vienna's greatest lover. Schumpeter then liked to tease his listeners by adding: I achieved two of them; the part about the top horseman didn't work out, unfortunately. But seen objectively, perhaps Schumpeter's greatest achievement is the fact that he is admired by liberals and Marxists alike for the incisiveness and complexity of his analysis of industrial capitalism. And no other great economist of the twentieth century comes close to his flamboyant biography. Schumpeter's dazzling career could have filled numerous ordinary lives and more.

Joseph Schumpeter started out as a child prodigy. Early in his career he authored a widely used book about the "essence of economics." After World War I, and for a short time, he was Austria's finance minister. Later, as an investment banker, he amassed a fortune, leveraged his speculative ventures, and ended this chapter of his life with a spectacular personal bankruptcy at the onset of a global economic crisis in 1924. Schumpeter then worked as a lawyer and government advisor in Cairo and Berlin, lived as a gentleman in London, and finally became the great explainer of the market economy's internal tensions and the role of creative entrepreneurship in economic progress. He was the author of texts shot full of insights over thousands of pages and held academic positions in Graz, Bonn, and finally, a year before the Nazis took power in Germany, at Harvard. Schumpeter's life was an uninterrupted sequence of creative destruction. Small in stature but with a towering intellect, irrepressible energy, and a thirst for recognition, with a longing for luxury and a delight in extravagance, a hedonist and manic worker at the same time, Schumpeter reinvented himself in every phase of his life, in sync with the central theme of his work: progress through innovation.

Capitalism Projects into the Future

Schumpeter was a child of his time, even more so than other great thinkers. Up until the 1930s, his view of the expansionary logic of capitalism was fundamentally optimistic and directed toward a better future, despite—or perhaps precisely because of—the era's brutal upheavals. Against classical economists obsessed with thinking in terms of equilibria, Schumpeter held up the creative power of the entrepreneur who, like an artist, constantly calls into question what is and fundamentally thinks in the future tense. For Schumpeter, stable capitalism is an oxymoron. The essence of capitalism is change, "discontinuities and seismic convulsions." That's the price society has to pay for waves of entrepreneurial innovation that bring progress through better and more affordable products, generate wealth that can be distributed, and finance the welfare state. In return, one had to accept major setbacks like the French financial crisis of 1924 that robbed Schumpeter of his fortune. Even after the New York stock market crash of 1929 and throughout the subsequent global economic cataclysm, Schumpeter remained convinced, contrary to the zeitgeist, that entrepreneurs' creative power, not counter-cyclical economic stimulus programs or monetary policy, would lead the world out of its economic upheaval toward renewed prosperity. He told his Harvard students that the Great Depression of the 1930s was a "cold douche" (apparently unaware of the word's double-meaning) for the economy that would spur many entrepreneurs to be even more innovative. Here Schumpeter's ideas were at complete odds with other schools of economics of his time.

In most write-ups, Schumpeter is called an "economist." That's correct—and at the same time imprecise, especially since Schumpeter is often mentioned together with leading figures of political economy of his time, from the classical economists to the neoclassical ones, demand theorists associated with John Maynard Keynes or monetarists like Milton Friedman. But at least in the public discussion of economic policy, it's often overlooked and strangely so that Schumpeter's perspective on the economy was at heart a microeconomic one.

Adam Smith, Keynes, and Friedman looked at the world of economics like a giant machine. This giant machine has control levers that can and must be adjusted on a macroeconomic level, whether through governmental or nongovernmental action. Schumpeter, on the other hand, imagined the world of economics starting with the individual, from inside out, and with the entrepreneur at the center, contemplating under what conditions these individuals can be productively creative for the long-term benefit of all—and under what conditions they can't. We will come back to this difference in greater detail in the next chapter. For us, it's the key to understanding how societies can deal more intelligently and constructively with rising waves of digital-economic change in our time. The issue of historically unprecedented concentration of power in digital markets looms large here. The question of market concentration is posed in a different manner now than at the time when Schumpeter, observing the world from Harvard, was still largely optimistic about capitalism's progress.

Capitalism Is about Capital

Through his microeconomic lens, Schumpeter saw no fundamental danger from monopolies born of economic dynamism. If an innovator succeeded in creating a *de facto* monopoly through radically better and more affordable products, Schumpeter didn't think the problem would be permanent. Sooner or later, these monopolies would be swept aside by the creative destruction brought about by new innovators, provided three conditions were met:

First, politicians must not let his adversary, John Maynard Keynes, mislead them into hobbling the power of competition and creative entrepreneurship by treating the economic system as fundamentally rigid and undermining its power for innovation through attempts at redistribution at the expense of (higher) national debt.

Second, a growing class of anti-capitalist intellectuals must be kept in check. These intellectuals do quite well for themselves materially under capitalism as scholars at universities, journalists, or functionaries for nonprofits, trade unions, and associations, and often also as public officials paid for by the taxpayer. But their business model consists of criticizing their host.

And third, in a functioning capitalist system, entrepreneurs must have access to the factor of production that gave the system its name: capital. Entrepreneurs need capital to finance innovation, that is, to make an idea marketable. Without capital for start-ups, according to Schumpeter, there can be no innovative economic dynamism because it's primarily entrepreneurs, not big traditional companies, that cater to innovation.

Beginning in 1911, in one of his early publications, Schumpeter emphasized the special role of the investor when it comes to start-ups. Ideally, entrepreneurs would simply obtain the necessary capital on the lending market in the form of long-term credits at market conditions. But in practice this often fails due to information asymmetries. Entrepreneurs don't know under what conditions capital is readily available. In Schumpeter's view, capital markets are too diverse, too complex, and too opaque. At the same time, investors don't know enough about the individual entrepreneurs and this drives up the cost of capital through unnecessary risk margins. That's why entrepreneurs often need an investor who not only provides capital but also promotes their cause and, in return, is rewarded with not just interest but a share of the business.

At the end of the 1930s, after enduring personal misfortune and against the backdrop of World War II, Schumpeter took a darker view of the market economy's logic of progressive development. His late work *Capitalism, Socialism and Democracy* in 1942 distilled his pessimistic view into a dire prediction: the decline of capitalism as a self-destructive force for change could no longer be stopped and the market economy would soon be replaced by socialism. Schumpeter feared innovation might be concentrated in a few large companies that offered their employees the best conditions of employment. And political pressure would then create an increasingly planned, decreasingly disruptive economy.

As we know today with the hindsight of history, the controversial contrarian had thought through the flaws of the market economy, summed them up with stylistic brilliance—and was nevertheless fundamentally wrong in his bleak prognosis. In the decades after his death in 1950, the governments of the Western world, especially the United States, recognized the capacity for innovation as the decisive competitive advantage in the struggle between East and West and did not systematically block its progress, as Schumpeter feared they would.

In the second half of the twentieth century, companies may have been dominated by paid executives instead of entrepreneurial personalities with an inner drive to create something fundamentally new, who stood high in Schumpeterian esteem. Bank loan officers may have looked first at what they stood to lose. Nevertheless, one of Schumpeter's greatest fears did not materialize. After World War II, the conjurers of creative destruction had easier access to capital than ever before in economic history. Capitalism did not destroy itself by suppressing creativity.

During the postwar years, the US government financed the development of new technologies—semiconductors, communications technology, space travel—as tools to win the Cold War. These technologies reinforced each other and laid the foundation for the information and telecommunications revolution. In Europe, thanks to the Marshall Plan, banks granted businesses generous loans that were often additionally secured by the government. Above all, a new class of capital assets emerged in the form of venture capital, which Schumpeter had named in an essay in 1943. And ironically, a colleague of his at Harvard, who had arrived in the

United States from Europe just a few years before him and became a professor if not in economics then at least in business administration, transferred Schumpeter's theory into financial practice: Georges Doriot.

Greed Wins

When it comes to the variety of roles and to lifetime intellectual accomplishments, Doriot, born in Paris in 1899 as the son of an automobile pioneer and race car driver, can't fully hold his own with Schumpeter. Georges Doriot's creative energy for bringing new things into the world at the intersection of science, teaching, and business was nevertheless impressive. In 1930 as a young Harvard professor, he helped found the *Centre de Perfectionnement aux Affaires* in the country of his birth. Today it's part of the elite HEC business school in Paris and offers one of the oldest MBA programs in the world. As a lucrative part-time job, Doriot represented Worms, a French commercial bank, in the United States and Canada. He became a US citizen during World War II and undertook a second career as a military strategist and planner, rising to the rank of brigadier general. At the time, some US spy agencies suspected him of espionage for the Nazis and the collaborationist Vichy regime, as well as money laundering for Worms. But no documentary proof ever emerged.

After the war, Doriot resumed his teaching appointment at Harvard Business School, remained a consultant to the Pentagon, and founded INSEAD business school in Paris, which evolved into one of Europe's most important management training center for up-and-coming executives. However, Doriot had his greatest impact via a much less famous organization. Shortly after returning to Harvard, he founded the American Research and Development Corporation (ARDC) together with physicist and former MIT president Karl Compton and two other partners.

ARDC was initially supposed to be a small investment company to provide soldiers returning from war with capital for starting their own business. The company enjoyed solid growth for a few years, almost entirely out of the public eye. That isn't surprising. The idea may have already existed for investors to enter into innovative companies at an early stage—but until then it had been more widespread on Broadway in financing new shows. Investors contributed start-up capital in order to put themselves in the limelight of high society, but above all in the hope of landing a hit show. Besides, in a postwar America of booming suburbs and surging consumer goods industries, there was little interest in funding technological disruption.

That changed in the 1950s due to a series of fortunate developments. In 1948, the physicist William Shockley and his colleagues at Bell Labs developed the first transistor, the central component not just in computers but in all integrated circuits. A short time later, Shockley went into business and hired some of the best minds in the field. For family reasons, he moved his company to Mountain View, California, establishing Silicon Valley. After receiving the Nobel Prize in Physics in 1956, however, he became more arrogant and aloof. Eight of his closest collaborators, the "Shockley Eight," parted their ways with him in 1957 and founded Fairchild Semiconductors, one of the first computer start-ups in Northern California. That same year, Doriot invested $70,000 of his own money through ARDC in an East Coast high-tech start-up called Digital Equipment Corporation (DEC). DEC went public in 1968 valued at $35.5 million. Doriot's investment had grown by a factor of 500, sending two messages to investors around the world with an appetite for risk: The future lay in high-tech. And there were big bets to be won. In contrast to young Joseph Schumpeter and his career as an investment banker, the elderly start-up financier Doriot didn't gamble away his profits. Rather, the investment in DEC was the first high point of a dynamic that would have delighted Schumpeter, but one he didn't see coming.

The same year DEC made Doriot rich, two Fairchild founders left the company and took one of their colleagues with them. They established a new chip start-up called INTEL. The money for it was raised by Eugene Kleiner, another Fairchild co-founder. Kleiner benefited from "founder's profit," as Schumpeter called it, and made a lucrative career out of his knack for investing in start-ups and guiding them to success. In 1972, Kleiner started the venture capital firm Kleiner Perkins, which financed many of the digital superstars of the decades thereafter, from Sun to Google and Amazon. Along with a number of other venture capital firms, Kleiner Perkins had sensed the emergence of massive business opportunities in high-tech fueled by creative entrepreneurship.

Since the 1960s, capital has been available in abundance for creative entrepreneurs in the United States, at least when entrepreneurs peddled the promise of transforming economy and society with the help of digital technology. Easy access to capital from investors with an appetite for risk enabled all the fundamental shifts in IT over the last fifty years: the establishment of the mainframe as a standard tool in large corporations, the development of microchips, the PC revolution, client-server architectures and the software revolution of the 1980s, the networking of computers locally and through the Internet, and the associated rise of e-commerce, search engines, online advertising, digital media, and online gaming. Venture capitalists financed social networks and social media, the simultaneous spread of mobile Internet, smartphones, and the associated app economy, online payments, and on-demand mobility providers.

No, Schumpeter's nightmare of the creeping self-destruction of capitalism through entrepreneurs' lack of access to sufficient capital did not come to pass. Even left-wing intellectuals, despite persistent attempts since 1968, have failed to overthrow capitalism. Instead, they rose through the institutions and have grown old as social democrats. And overly greedy managers hungry for their next quarterly bonus didn't prevent Western and Japanese-style market economies from generating enough innovation, growth, and prosperity to expose Eastern European planned economies as hidebound pseudo-alternatives (which, measured by key economic indicators, they were until their self-abolition in 1989). One does not even have to factor in the lack of freedom under socialism to see that.

The New Nightmare

If Joseph Schumpeter were alive today and could observe the economic conditions of the digital world with his keen historical acumen, he would have rather different worries. He would likely still consider it a nightmare, just a fundamentally different one than the one he had feared. Schumpeter would realize that market concentration and quasi-monopolies in advanced digital markets did not turn out to be transitional phenomena, as Netscape, Nokia, and MySpace once were, but solidified in a way that's both remarkable and alarming. The Apple/Google duopoly in smartphone operating systems and their connected app stores—a triumph for both companies—is only the most obvious example.

Microsoft has found its way back to dominance of PC operating systems and office software, now complemented with a place in the cloud computing oligopoly. The acquisitions of Skype and LinkedIn have also brought new power to the old empire of Bill Gates. The Facebook conglomerate, including Facebook (2.6 billion users), WhatsApp (2 billion users), and Instagram (1 billion users), connects a significant portion of all Internet users worldwide, and, together with Google, takes in around half of all online advertising revenues. With YouTube, Google has conquered the market among open video platforms, allowing just about enough space for niche providers like Vimeo. Amazon already has about 50 percent market share in online retail in the United States, which it is growing through its premium Amazon Prime service. Almost incidentally, Prime has in turn made Amazon the most successful streaming service for TV series and movies after Netflix. At least, this streaming duopoly is now facing a challenge from Disney, but Schumpeter would certainly remain suspicious of such three-way market diversity.

With other digital platforms covering on-demand mobility and short-term rentals, travel reservation and dating, food delivery and price comparisons, jobs and freelancing, online gaming and gambling, the same network effects and forces at work promoting concentration of power produce the same result: winners take (most of) all. Often, a few new providers create at least a little competition through smart use of market follower strategies. But the forces of creative destruction are rarely seen in these markets, and if they are, these start-ups quickly find themselves in the so-called "kill zone" of the digital power players. "Kill zone" is where the Goliaths simply buy up the Davids. It happens when venture capital–funded start-ups push their innovative ideas in ways that threaten digital superstars, as occurred with WhatsApp and Instagram, Wunderlist and YouTube, Skype and Farecast. It's more often than not extremely lucrative for start-up founders, albeit frustrating as they lose the chance to develop their product the way they had imagined. If founders refuse the offer, superstars will often copy the key innovative features and entice users who are already tied to their systems, leaving challengers high and dry. If that happens, even more venture capital can't save the Davids.

In the start-up world, things seem to be going increasingly the way soccer romantics wish their sport was: money doesn't score goals. With soccer, sometimes they're right. In the world of venture capital, on the other hand, increasing amounts of start-up capital rarely translates into victory against superstars. The series of failures of Softbank's venture funds in recent years is evidence that successes of large platforms can neither be reproduced with vast amounts of money and copycat models, nor can they be disrupted with equally vast amounts of money and fundamentally new ideas.

That might have astonished Doriot even more than Schumpeter. Venture capital was a fundamental innovation in the capital market. It enabled creative minds to explore expensive and risky ideas and turn them into innovations in the Schumpeterian sense: new products that are successful in an established market or, even better, in an emerging one. Traditional banks, trapped by legal requirements, ingrained patterns of thought, and their owners' expectations of stable returns, could not (and cannot) place risky bets on disruptive innovations. ARDC and other pioneers of venture capital solved a financing problem at a time when it was expensive to reinvent the world with microelectronics and no one else was willing to take the risk.

For venture capitalists, of course, this innovation in capital markets turned out to be a highly profitable business. In retrospect, however, the innovative effect of this achievement has been under- not overestimated. From the late 1950s onwards, venture capital made money, a scarce resource in those days, accessible to the most innovative minds. This reallocation of capital enabled leaps in technological innovation that in turn paved the way for the greatest aggregate transformation of the West since World War II: the shift from industrial to knowledge-based societies based on information technologies. The futurologist Alvin Toffler summed up the socio-psychological consequences of the changes caused by computers and large-scale technologies in his 1970 book *Future Shock*. Too much change in too short a time overwhelms individuals and societies. Toffler did more than describe a prevalent feeling at the time. He described facts: change could be measured by, among other things, a steeply rising rate of innovation—that is, to employ but one measure, the share of sales from products less than three years old.

Technology's Frenzied Standstill

If Schumpeter, Doriot, and the Shockley Eight together could observe the course of technology and the digital economy in the present, they would probably agree: People today only believe they're living in incredibly innovative times. They assure each other that one digital innovation is followed by the next, revolutionizing the way we work and live at breakneck speed. But these people—that is to say, we—suffer from the vanity of the Here and

Now. We deceive ourselves, pretending to live in an era of frantic technological change, at a tipping point in world history, as life is fundamentally reshaped by technology. It makes us feel important. After all, who wants to live in boring times? But in fact, ours is a time of comparatively scant innovation.

Joseph Schumpeter lived from 1883 to 1950. During this time, modern toilets made their way into almost every household, radically improving hygiene. Ocean liners, trains, and bicycles became means of mass transport. The automobile was born, along with the telephone, radio, and movies. X-ray machines revolutionized the medical profession and antibiotics became the most powerful medicine of all time. Artificial fertilizers drove the green revolution. As a student, Schumpeter read about the Wright Brothers' first powered flight in the newspaper. By the time he became a professor at Harvard, there was already an aviation industry. In the decade before his death, computers and nuclear power entered the world. He narrowly missed the first human being in space. Georges Doriot, for his part, died in 1987. At the time, Douglas Engelbart's computer mouse was already twenty-five years old, the first Mac hopelessly outdated, robots were working in factories all over the world, and the TCP/IP protocol, the basis for worldwide computer networking, had been established. Soon, the British physicist and computer scientist Tim Berners-Lee would be working on making access to information through the World Wide Web even more convenient. In comparison, what level of innovation do the iPhone and Alexa, Zoom conferences and scooter sharing, Tinder and Google Now represent? Breathless cycles of technology hype and an associated buzzword bingo remind us of Paul Virilio's metaphor of "polar inertia." We live in a time of such frenzied standstill in innovation.

Anyone who describes this phenomenon at tech conferences, for example by referring to the work of economists Tyler Cowen and Robert Gordon on decreasing productivity growth despite the supposedly high rates of innovation, is often met with surprising aggression. Yet a sober look at the relevant numbers shows: we live in boring times. In comparison to the last 150 years, the rate of innovation today is alarmingly low. A glance at a range of economic indicators confirms that impression of frenzied standstill. In the United States, as in the Western world more generally, productivity gains are at historically low levels. The number of new US companies within total corporate numbers is less than half that of 1980. And a shrinking number of companies are registering the majority of patents—a further indicator of the economy's flagging power for innovation. Comprehensive analyses show that almost all regions and economic sectors in the United States are affected by this slowdown in the dynamism of innovation. By some indicators, initially high-tech seemed to be different; but research, for instance by Ryan Decker and his colleagues, shows that it has lost business dynamism quite pronouncedly since the early 2000s.

In recent years, experts such as NYU economist Thomas Philippon have pointed out that an important cause of the decline in dynamism and innovation is increased market concentration. This is not only true of the digital superstar companies that dominate their respective markets. Even in banking and financial services, among airlines and telecommunications providers, competition is falling, and prices are rising. Just as US universities have increased tuition fees over the past two decades like a cartel, the lack of competition in health care and pharmaceuticals is now chronic. Philippon and many of his colleagues blame US antitrust regulators, who are neglecting their task of ensuring market diversity and robust competition and preventing the abuse of market power. The failure of government authorities is obvious. But at best it's only part of the explanation.

US economic data on increasing market concentration of course reflects the rise of digital superstar companies with their astronomical market capitalizations, rates of return, and market shares. If they were as creative and innovative as they and many tech pundits claim, increased market concentration wouldn't come at the cost of decreased capacity to innovate. On the contrary, there would be an ever-increasing stream of things being hatched and at an accelerating pace. Setting any vanity of the Here and Now aside, hard data on market concentration, declining innovation rates, stagnating productivity, and increasing rates of return earned by large corporations in general and digital superstar companies in particular leads to one conclusion: As customers and tax-paying citizens, we are being doubly cheated out of the benefits that technology-based innovation has provided since industrialization. Progress is slower than under healthy competition. And large corporations also enjoy monopoly- or oligopoly-level returns on investment at the expense of customers and society.

Under conditions described by Schumpeter, none of this should be happening, or, at most, be little more than a brief transitional phenomenon. In numerous start-up hubs around the world, from Shenzhen, Bangalore, and Barcelona to New York, Houston, and San Francisco, there are enough smart entrepreneurs with easy access to venture capital. But, unfortunately, in the last few years they've had much less impact than their hyped business valuations imply, perhaps following the start-up meme of "fake it 'til you make it."

In advanced data capitalism, ideas and money are no longer enough for creative destruction. As shown in the previous chapter, even affordable computing power, open source tools, and free algorithms help only a little, if at all. The real scarce resource of our boring times is digital information. While data liquidity alone may not be sufficient, it's a necessary prerequisite for innovative entrepreneurship in almost all industries and contexts. The companies with the greatest abundance of data they use tend to be the most innovative. But even that innovative dynamism of digital superstars is limited. And through an innovative analysis of data, Boston College economist Simcha Barkai found that larger firms, including in High Tech, are able to charge higher prices than under full competition. So we all pay the price, quite literally, as the lack of competition doesn't force Big Tech to think hard, and exclusive access to data is like a license to print money—a lot of it.

To make matters worse, it's easier for superstars to attract and retain creative and highly educated talent. Of course, this is related to the high salaries the data giants can pay. But in the future, talented young people with creative ambitions may also increasingly realize that they can only do truly creative work if they have access to data. More generally, who wants to kick their heels at companies with no future? As a consequence, companies lacking access to data will first run out of talent, then ideas, and finally customers and money.

The asymmetry of information and power due to data monopolies would make Schumpeter furious. He might even risk his life to get rid of it. That may sound melodramatic, but as Schumpeter's biographers report, that's precisely what he did as a young professor in order to get his students free access to knowledge. A few weeks after starting his teaching position at the University of Czernowitz, Schumpeter presented his students with an extensive reading list. A boorish librarian, however, restricted loaning the books to students, although the library owned plenty of copies. In response, Schumpeter challenged the librarian to a fencing duel. The librarian left the dueling ground with a deep cut on his shoulder and granted free access.

4. *Data Capitalism*

Every innovation begins with an idea. In the mid-1960s, Ralph Henry Baer had a big idea for a small, brown box. The idea grew into an industry that's a textbook example of Schumpeterian innovation.

Baer was born in 1922 in Germany's southwestern Palatinate region. In 1938, two months before Kristallnacht, his Jewish family fled to the United States. He trained as a radio technician. During World War II, Baer was stationed in London with a US Army intelligence unit. After the war, he studied a subject with a future: television engineering. For most of his professional career, he developed electronics for the defense firm Sanders Associates. On the side, Ralph Baer thought about what TVs could also be used for. It occurred to him that the early computer games for mainframes required little computing power, even by the limited standards of the time. Couldn't chips be developed to turn TVs into gaming computers? In 1966, Baer convinced his bosses to invest $2,500 in his idea. In 1968, after two years of development, he presented the "Brown Box," the world's first video game console.

The Sanders management regarded this invention as a "waste of the company's time." They didn't see any business opportunity for them, but they did patent the invention. Baer looked for other partners. Four years later, the small brown box appeared as a licensed product under the name "Magnavox Odyssey." It quickly became clear that even toys can create new markets with disruptive technology, especially if these toys are not only fun for children, but also trigger addictive behavior in adolescents and young adults. Baer was soon given the unofficial title of "Father of Video Games." Many years later, he received the National Medal of Technology from President George W. Bush.

In the meantime, the new market for game consoles turned into a petri dish for economists who could observe how, in cyclical waves, new market players challenge the industry's established top dogs with technological innovations and push them aside. The newcomers would then dominate the market, only to be disrupted again a few years later by other platforms with better devices and more interesting games. Magnavox was forced to give way to Atari. "Pong" ran more smoothly than Baer's "table tennis." Fairchild, Intellivision, and Coleco vigorously shook up the market in the following years, but Atari was ultimately pushed aside by a Japanese competitor, Nintendo, and its superstar Super Mario. Another newcomer, Sega, in turn made life difficult for Nintendo. But then Sony brought to market its PlayStation and Sega declined. Out of nowhere, suddenly Microsoft appeared on the scene with the Xbox and made Sony look dated. The Japanese countered with the Wii, a cross-generational hit. In an important study, Mark Lemley and Andrew McCreary of Stanford Law School used the history of the gaming console industry as an instructive example of something that rarely happens in the digital economy today: innovation by new actors in the market.

The US digital top dogs, Apple and Microsoft, are more than forty years old. Google, Amazon, eBay, and Netflix were founded in the last century. The latest generation of digital innovators—Facebook, Twitter, Uber, and Airbnb—were born in the aughts. There is little sign of cyclical renewal by new market participants in important digital markets today. Why? Lemley and McCreary distill the answer from their ninety-page article into just two eponymous words: "Exit Strategy."

The Exit Mantra

In the previous chapter we described the concept of the "kill zone" that start-up founders can sometimes stumble into when their development takes them too close to the business model of superstar companies. "Kill zone" makes it sound like the start-ups are the victim, but that's only half the truth. The authors of "Exit Strategy" rightly ask why founders think so intensively about their strategy for getting out even before starting a company, when they are still sketching their idea and drawing up a business plan, recruiting the first angel investors and developing the celebrated "minimal viable product." One would expect an entrepreneur to pursue the goal of owning and managing the company as long as possible. In the nineteenth and twentieth centuries founders often subscribed to a biological exit strategy; they bequeathed their company to their descendants. The management mantra of the entrepreneurs in Silicon Valley, on the other hand, is no longer to "disrupt," but rather to "cash out." That may appear a shade too strong, but Lemley and McCreary offer some interesting numbers to support the idea.

Traditionally, (successful) entrepreneurs and venture capitalists have had two options for turning their investment of energy, time, and capital into (a lot of) money. They can either try to go public or sell some or all of the equity to another, mostly larger, company. After an exit via IPO, founders usually remain involved. This was the case with all superstar companies mentioned earlier, from Apple to Airbnb. It was also the express goal of the founders, and obviously this path was also supported (at least for a time) by the venture capitalists. In the 1990s, as Lemley and McCreary meticulously pull together, half of all (commercially successful) companies went public when venture capitalists had financed their rise. In recent years, in contrast, it has been less than 10 percent. Nine out of ten successful founders opt for an exit involving a generous acquisition offer from an established technology company with deep pockets.

As of mid-2021, Google has taken over more than 260 companies since its early days, Facebook almost 130, Apple about 125 and Amazon around 110. Presumably the venture capitalists involved popped open bottles of champagne to celebrate. One founder or the other may have sold his shares to Big Tech with mixed feelings, but most of them probably were well aware of the game they had gotten themselves into. Their job as entrepreneurs was to originate an innovation that could threaten Big Tech. If they succeeded, the so-called "kill zone" wasn't a place for killers and victims, but rather a place where money got handed over. The founders left the zone as multimillionaires, sometimes even billionaires, but weren't allowed to start anything similar. Big tech companies, as they always emphasize in their press releases on the takeovers, not only buy technology that's allegedly a "perfect" complement to their existing portfolio, but also acquire numerous talented employees in the process. Such press releases, however, usually omit the one sentence that would more aptly describe this: "We took over a potential competitor and were prepared to pay a price well above market value for it."

Facebook has always been particularly generous when it comes to paying extra for potential competitors. In 2012 it shelled out around $1 billion for Instagram, still a small photo community at the time. Two years later it paid $22 billion for the message service WhatsApp. At the time, observers considered both amounts to be staggeringly excessive. Facebook founder Mark Zuckerberg, on the other hand, saw not only opportunities for the parent company in the acquisitions, but also existential danger: Instagram allowed users to share photos simultaneously across multiple platforms, including Facebook and Twitter. Use of this feature was growing rapidly, potentially endangering Facebook's dominance.

Facebook had also missed the growth of message services. Perhaps these services would develop into social media platforms with their own network effects. The astronomical price Facebook paid for WhatsApp stemmed from a simple calculation: fending off a potentially existential threat. The actual price didn't matter that much, especially since money was piling up in Facebook's accounts thanks to monopoly-level returns on investment. But Facebook and other Big Tech companies drew a rational conclusion for the following years: it's better to buy competitors sooner than later or else it'll cost far too much.

Of its publicly declared $25 billion for start-up acquisitions in the decade since 2008, Google has spent more than 75 percent on direct competitors. The most prominent example is the Waze map service, which threatened to overthrow the dominance of Google Maps. In recent years, Apple has tried to meet the challenge from its new competitor Spotify by buying up high-tech music companies such as the cult headphones manufacturer Beats and music recognition app Shazam. Under the leadership of Satya Nadella, the relatively venerable Microsoft re-entered the bidding race for hot start-ups, or perhaps more precisely for so-called "grown-ups," that is, successfully established start-ups with a more or less functioning business model and solid customer relationships. The billions invested in LinkedIn, GitHub, Minecraft-creator Mojang, and speech-recognition leader Nuance seem to be paying off for the now middle-aged Microsoft. This astute M&A strategy is considered to be one of the factors driving the rise of its stock price and making it once again one of the world's most valuable companies.

These new roles in the race for capturing digital innovation show that many entrepreneurs still aim to revolutionize the world through technology. However, given the speed of digital development cycles and the prevailing competitive pressures, this can't be done without venture capital. And anyone who wants to make a deal with a venture capitalist needs a good exit strategy, because capitalists only take big risks if there's a chance for a big quick payoff. Today the likelihood of selling out to Big Tech is ten times as high as going public. It's also sometimes more lucrative as tech giants are willing to pay a premium. Mark Lemley and Andrew McCreary conclude: "A culture of vibrant startups that should drive Schumpeterian competition by leapfrogging less nimble incumbents has been coopted by the structure of the VC market." The established companies are paying exorbitant prices for technology they won't even use for one simple reason: "Eliminating potential competitors keeps their profits high."

In the kill zone, it's not the start-ups that are dying, but Schumpeterian innovation, at the cost of market concentration and monopoly formation. This billion-dollar game to the detriment of customers and our future is not all that hard to suss out. Especially since one of Silicon Valley's most prominent venture capitalists wrote the instruction manual for it in a book that topped the *New York Times* bestseller list in 2014.

Creative Monopolies?

"Competition is for losers." Peter Thiel offered this judgment in numerous interviews about his book *Zero to One*. And then he added a clarification: "If you want to create and capture lasting value, look to build a monopoly." From a digital entrepreneur's point of view, this makes a lot of sense. Why should the founder of a social network, a movie streaming service, or a digital payment system willingly accept the same fierce competition that is faced by retailers, airlines, or automotive manufacturers?

Thiel is a trained lawyer. He used his money, his contacts, and his skills as a venture capitalist to help PayPal, Facebook, and leading analytics firm Palantir achieve oligopoly status in their markets. That should already be enough to earn a spot in Silicon Valley's Hall of Fame. But the conservative libertarian maverick and Trump supporter, chess master and philosophy aficionado, hard-nosed negotiator and gadfly, is widely unpopular in Silicon Valley. Perhaps it's because he doesn't refrain from pointing out the digital economy's ugly white elephant in the room.

In 2012, Thiel took part in a panel discussion at the Aspen Institute in Colorado with Google's then executive chairman Eric Schmidt. Schmidt told the usual story about Silicon Valley's outstanding capacity for innovation thanks to its rich and diverse start-up scene and unique ecosystem of large and small players who single-mindedly improve the world through digital progress. In short, Schmidt told the story that Silicon Valley loves to tell and that many technological latecomers continue to believe to this day. In response, Thiel called the Google executive a "minister of propaganda." Schmidt was visibly not amused. For several years, Peter Thiel has been living in Los Angeles from where he has accused entrepreneurs and superstar companies in Northern California of having lost the capability for creative destruction. They suffer from "conformity of thought." Silicon Valley is no longer attracting talented people, he suggests, "but the greedy ones."

Thiel's tone is often disconcerting. So is the matter-of-factness with which he sums up the philosophy of his time in Silicon Valley that made him personally rich and powerful: big digital companies search for unlimited economic power, pushing others to the wall, forging smart alliances with major players, and taking over competitors until finally only one winner remains. Channeling Schumpeter, Thiel calls these winners "creative monopolies," which for him has entirely positive connotations. For Thiel, creating such a monopoly is and remains the goal of digital start-ups because a "creative monopoly means new products that benefit everybody and sustainable profits for the creator." Competition, however, means "no profits for anybody, no meaningful differentiation, and a struggle for survival."

Many CEOs of digital superstars would probably rather bite their tongues than publicly agree with Thiel. But the companies they run are successful cases of "creative monopoly formation." And with their current acquisition strategies, the CEOs are ensuring no new "creative monopolies" arise. What we find particularly interesting is that the attitude and *modus operandi* within Silicon Valley's culture are very similar to a system that is often presented by tech-enthused Americans as its diametric opposite. Drawing on the works of MIT economist Yasheng Huang and others, we're thinking of the Chinese variety of digital monopoly capitalism.

Digital Capitalism with Chinese Characteristics

Half admiring, half intimidated, the West has watched China's rise to digital superpower status. Often our attention focuses on the role of government. Through the Great Chinese Firewall and economic protectionism, policy makers in Beijing have shielded Chinese search engines, e-commerce platforms, and social network start-ups from foreign competition in a huge domestic market. With Baidu, Alibaba, and Tencent (the operator of WeChat), companies emerged whose size and profitability put them in the same league as US digital superstars and whose applications in some areas bested their US counterparts in terms of size as well as innovativeness. Bit by bit, platforms as robust as those in the United States took shape in almost all significant areas of the digital economy, including Uber clone Didi Chuxing, Netflix-like streaming services iQiyi (from Baidu) and YY, and, of course, the mobile payment platforms that turned China into an almost cashless economy. In the meantime, Huawei joined Apple and Samsung in the top league of smartphone manufacturers.

Wherever hardware development plays an important role, Chinese companies are now often technologically superior. European companies like Nokia and Ericsson can't compete on price or quality with Huawei's 5G-network infrastructure technology, and there are no American providers. For drones, DJI is the clear global leader in both innovation and market share, while in Alibaba's futuristic Freshippo grocery stores, robots politely and efficiently serve customers. Reports from this new land of unlimited digital possibilities are certainly exaggerated at times. The logic of the hype cycle also applies in the Far East. There, too, the prototypes, first applications, and grand promises can be followed occasionally by phases of disillusionment. But digital start-ups and the Chinese government have co-written an impressive success story over the last decade. And the roles between the two are far more cleverly devised than is often perceived in the United States and Europe.

The West often harbors the notion that Chinese government technocrats pick technologies and applications they deem important for the country's development, then select entrepreneurs and start-ups to turn them into national champions by providing capital, market privileges, and government contracts until they're fat and powerful enough to conquer the global market. This strategy was used until well into the aughts, primarily with state-owned companies (SOEs). Anyone taking a look at the central government's big strategy papers, such as the high-tech initiative "Made in China 2025" or the "Internet Plus" development plan for mobile Internet, Big Data, cloud computing, and the Internet of Things. might perhaps believe that this was still the case. But today's reality of creative entrepreneurship in China is a bit different.

Chinese policy makers now understand that only the framework conditions for innovation can be planned, not innovation itself. In many fields of technology, the Chinese government is acting much like Silicon Valley venture capitalists did in the 1990s and 2000s. They examine important technology trends, make available capital relatively broadly through state banks or semi-governmental venture funds, and watch which start-ups and services prevail in the competition of ideas and their implementation.

At this point, a significant portion of the capital used in scaling up is private. In general, Chinese authorities don't prevent competition, takeovers, or bankruptcies. Rather, the opposite. The decision makers in Beijing are increasingly following the original Silicon Valley model that Schumpeter had sketched out. They see competition as the driving force of digital innovation, and, as in the West, they don't stop the long march of successful start-ups toward market dominance.

Government has been intervening to stop unwanted political content, and to maintain access to data for surveillance and control purposes, especially on the dominating digital platforms operated by Baidu, Alibaba, and Tencent. And these companies have been cooperating with government agencies to an extent that would be shocking in Western democracies. But the government's access to data is driven by the Communist party's iron will to ensure political stability and the party's monopoly on political power; it is mostly political rather than economic. In terms of entrepreneurship and start-up culture, the situation in China is surprisingly similar to that in the United States. Venture capital, network effects, heightened customer focus, good logistics, and, of course, entrepreneurial skills, energy, and charisma paved the way for a number of "creative monopolies," even without major government intervention.

In other areas of development, such as Huawei's network technology or SenseTime's surveillance AI, government did indeed stimulate demand through procurement contracts, but also relied on fierce competition among different start-ups. Only at a later stage have policy makers promoted market consolidation, facilitating the emergence of a similar market structure as in the United States: a few market-dominant superstars, who thanks to limited competition achieve high profit margins. Of course, national champions are also geopolitically important for the Chinese government. The first skirmishes in a so-called "Tech Cold War"—we'll explore the concept in more detail in the final chapter—hinted at how China plans to use technology exports as a geopolitical instrument of projecting power (economic and otherwise), while US administrations from Trump to Biden have been trying to defend American technological superiority by making access to American software and high-performance chips more difficult for Chinese companies, or by trying to ban Chinese apps from Google's and Apple's app stores, in effect closing off the Western market. The new battle lines of the two digital superpowers reveal how strongly their technical capabilities, ambitions, and policy approaches have come to converge.

Seen through the lens of political economy, an intriguing spectacle is unfolding: in the sphere of foreign trade, the United States is resorting to protectionism just as Chinese companies are challenging American technological dominance. The rise of Chinese tech corporations to superstar status with global operations was only possible because a Communist government instructed economic bureaucrats not to meddle too much with entrepreneurial development, permitting the logic of digital growth toward duopoly and quasi-monopoly to run its course. For example, Tenpay and Alipay together process more than 90 percent of smartphone payments in China. Digital capitalism with Chinese characteristics doesn't have a problem with monopolies. The Chinese government can decide to regulate, transform, or dissolve such monopolies at any time it so chooses (as exemplified by how it stopped the IPO of Alibaba's banking arm Ant in 2020). On the other hand, there's another parallel that's often overlooked—for years, government has refrained from regulating markets, not only in China, but also in the United States.

Throughout the last fifteen years, US federal antitrust authorities have been tame—not because they lack resources or the ability to act, but because of deliberate policy choices by the respective administrations. The last major attempt to put a dominant tech company in its place and to ensure fair competition was more than twenty years ago. In the 1990s, enforcers of antitrust law initially cracked down harshly on Microsoft, accusing it of abusing its market power by bundling its Internet Explorer browser with its operating system, and by letting its Office software division use special hooks into the Windows operating system unavailable to competitors. Legal appeals and political deals just about prevented the breakup initially ordered by a federal court in 2001.

During the next two decades, no Big Tech company faced a similar threat from US federal antitrust lawyers. Unlike their state counterparts, they were not engaging in comprehensive lawsuits and threatening huge fines. The situation was rather different on the other side of the Atlantic, thanks to a charismatic champion of market competition from Denmark who Time magazine ran a story about in May 2015 titled, "Why This Woman Is Google's Worst Nightmare."

The Danish Competition Advocate

Margrethe Vestager was born into a family of pastors with deep roots in the Danish liberal tradition. One of her great-great-grandfathers was a co-founder of the Danish liberal party. Vestager studied economics; early on she embraced market principles like competition, options, and free choice, and stood for them in other contexts as well. Due to her razor-sharp intellect, her rhetorical eloquence, and her natural empathy for the needs and problems of citizens, she rose quickly in Danish politics. After a stint as Danish minister of education, in 2014 Vestager was nominated to become EU Commissioner for competition. Among the commissioners in Brussels, the competition portfolio is arguably the most powerful. It is like the FTC and the Department of Justice's antitrust division folded into one, with a sprinkling of the SEC on top. From the very early days of the European Union, ensuring robust competition has been a core policy principle. As Margrethe Vestager took over, she vowed to exercise her power.

Vestager wasn't just targeting Silicon Valley Big Tech. She fought against a cartel among European truck manufacturers, criticized Russian energy giant Gazprom for abusing its market power in Eastern Europe, and opposed large telecom mergers. Mastercard was served a fine in excess of $600 million. But Vestager, who believes government needs to ensure and enforce that markets stay competitive, reserved her biggest punches for Big Tech. No politician around the world is pushing back harder against digital monopolists than her. During her reign Google alone, as we mentioned, was fined over $8 billion for anticompetitive behavior.

Of course, not everyone appreciates Vestager's fight for competitive markets. Big Tech, and Google in particular, with its vast army of lawyers appealed the EU Commission's fines. A more reluctant, and less market-oriented European Court of Justice repeatedly reversed judgment or reduced fines. In the 2018/19 elections for European Parliament, Margrethe Vestager ran for Commission president. When the large nations picked German veteran politician Ursula von der Leyen instead, some media wrote off Vestager. But instead of fading away, von der Leyen picked Vestager as executive vice president with a broadened portfolio of all things digital and let her retain the competition portfolio as well. Not only Google's but Big Tech's nightmare came true: rather than being elevated to the far more general Commission presidency, Vestager's new job was a signal by the EU member states that they wanted her not only to press on but double down on reining in data monopoly capitalism.

Vestager's role is powerful. Her directness in calling out digital monopolists made her popular throughout the continent. But important forces are pushing back—not just Big Tech, their lawyers and lobbyists, and a more reluctant, less tech savvy European Court. Stringent opposition also comes from a somewhat unlikely corner: Thierry Breton, the EU Commissioner for the Internal Market and former head of large (and sometimes dominant) French IT and telecom companies.

While Vestager combines her liberal pro-market views with European values to push for market competition, Breton aims to combine European values with old fashioned industrial policy, highlighting the importance of national and European champions. The tension between Vestager and Breton surfaces a deeper and more troubling incoherence in Europe's position to respond to Big Tech.

Moral Superiority and Digital Sovereignty

If a psychologist were to examine Europe's state of mind regarding its digital present and future, they would possibly diagnose a bipolar disorder. Phases of pessimism and depression alternate rapidly with ostentatious displays of optimism, political braggadocio, and action for the sake of action. What's the source of Europe's deeply disturbed digital emotions?

There is a fear that information technology at its core is a tool used by the powerful to control, direct, manipulate, and, in the worst case, oppress dependent and vulnerable people. In Europe's debates over digital policy, stories about the rise of foreign Big Tech companies are often related as tales of enemy espionage. The American version of this story is Edward Snowden's military-digital-industrial complex and Shoshana Zuboff's "Surveillance Capitalism," in which Apple, Google, Facebook, and other tech giants know ever more about us and thus make more and more out of us. Not surprisingly therefore, Zuboff's theses received huge attention in Europe. The Chinese version is the dystopian story of a reinvention of the surveillance state by digital means: a digitally administered Social Credit System, constant network censorship, and secretly installed surveillance apps that monitor every step we take, every connection we forge, every action we plan.

Europe tries to channel this deeply rooted fear into a moral discourse whose most important strand is digital and data ethics. It is laced with "European values." As important and useful as this discourse is, it's nevertheless striking how the chattering classes of academia, politics, and civil society share a sense of moral superiority when it comes to Silicon Valley's model of "surveillance capitalism" and even more so toward China's model of "digital authoritarianism."

There seems to be agreement that both models are no option for the old continent and that the protection of fundamental digital rights, in particular data protection, must steer us in the direction of an alternative European approach. This alternative must protect privacy and promote political participation while at the same time create economic prosperity for everyone. Data protection and democracy may play well together in the discussion, but the third dimension—growth, prosperity, economic power—is increasingly at odds with the current narrative of Europe's independent, better way to a digital future. Leading economic indicators offer no reason for euphoria. Real gross domestic product in the European Union in the decade before the pandemic has grown by just about 1.5 percent annually—about half the growth rate of the overall global economy. Europe's share of world economic output is decreasing and today stands at around just two-thirds of what it was twenty years ago. In significant part, this is because Europe is lagging behind when it comes to digital innovation.

European data privacy advocates have long tried to market data protection as a competitive advantage. They suggested the whole world would use European cloud services and applications, because data is safe there and nobody is reading over one's shoulder. Some proponents of the California Consumer Privacy Act have adopted this view and looked at European data protection legislation as a role model for their own regulatory initiative. But, having watched the European situation up close, we are skeptical. Not only are European data protection rules in practice little more than formal hoops consumers click through without actually being empowered (more on that later), they also failed to produce an economic dividend.

Europe has had in place stringent data protection laws since the 1990s (and not only since GDPR) that should have been reflected in economic growth and demand for European digital services. But even a cursory glance at the list of the ten most valuable digital companies in the world confronts the unpleasant reality that

European digital companies are nowhere near the size and importance of Silicon Valley Big Tech.

There are many good reasons for data protection but passing it off as a competitive advantage in today's global digital markets is either naive or dishonest. In digital policy debates, the fear of surveillance through digital technology is therefore increasingly meeting its dialectical counterpart: a simultaneous fear of economic decline due to insufficient value creation with and through digital technology. Nearly all debates under the rubric of "digital transformation" in recent years—whether for particular companies or for national economic policy—have been shaped by this fear of decline.

That is when Thierry Breton and his national counterparts come in, pushing for a a policy instrument used in the previous century: conventional industrial policy. That's when Europe's anxiety disorder shifts to a euphoric certainty that the "old continent" is on the verge of an unparalleled surge to catch up with and surpass the leaders of the global digital economy. And that this will not only bring new prosperity to the continent, but also generate Europe's "digital sovereignty." And the quoted precedent for this is the biggest success story of European industrial policy: Airbus.

A New IT Airbus?

At the end of the 1960s, Europeans recognized that it was only through cooperation that they would be able to develop and produce passenger aircraft that could take on American manufacturers Boeing and McDonnell Douglas. Through political coordination, financial subsidies, and support from state airlines in the form of firm orders, and after setbacks and successes, a remarkable company has in fact emerged. Today it not only builds outstanding aircraft but is also an inspiring example of the productive strength of European cooperation. Airbus has been the largest aircraft manufacturer in the world since the end of 2019 and has pulled ahead of Boeing, its troubled fellow duopolist. If backers in politics and industry have their way, European taxpayers will pay billions for a similar industrial policy to build a new European digital infrastructure, GAIA-X.

The project was spearheaded in France, the home of dirigiste economic policy, and Germany, the home of data protection and digital ethics. In 2020, then German minister for economic affairs Peter Altmaier called it "perhaps the most ambitious digital project of this decade" and a "digital moonshot." This earned him much applause from European research bodies, IT companies, and industry associations, as well as from data protection advocates and network policy makers. GAIA-X sounds like a smashing success, and makes the companies potentially charged with implementing it dream of astronomical profits. Just a glance at the companies involved makes clear that almost all the major economic players worth considering in France and Germany will soon be submitting big bills to European and national authorities.

GAIA-X was presented as a project to create a European competitor to American and Chinese cloud providers, focused on unifying different data standards. In principle that's not such a bad idea. More competition is always good, as Schumpeter reminded us. However, the idea of GAIA-X competing against AWS, Azure, and Alibaba cloud services raises questions: Is such conventional industrial policy to produce a European "champion" feasible, and at what cost? Will such a new entrant have a chance against the "hyperscalers," such as Amazon, Microsoft, and Alibaba? Should taxpayers shoulder a part of the financial risk for such a private sector undertaking?

Initially, 2 billion euros from funds set up to boost the economy in the wake of the pandemic were earmarked for starting GAIA-X. Most of the financial and economic policy makers involved are likely aware that this isn't nearly enough for a digital moonshot. It's at best enough to level the ground for the launch pad. Big Tech cloud providers each invested hundreds of billions of dollars in a conceptually far less ambitious infrastructure. How many euros from public coffers would it take to craft a European IT Airbus into a serious cloud competitor? Cloud computing is a highly commodified service. Amazon, Alibaba, Microsoft, and other Big Tech companies have the ability to operate their clouds in a mixed model that combines utilizing the infrastructure themselves and renting out excess capacity. Their storage and processing powers will continue to rise, and prices will continue to fall; only the most efficient providers on the market will be able to earn a profit.

And how plausible is the historical comparison with Airbus? Unlike building airplanes, technical innovation in cloud computing isn't incremental and linear, but leaps forward. Governments can't stimulate demand as they did through purchases from state-owned flag carriers, and competitors have had many years' head start, and amassed a mountain of operational experience, besides having extremely deep pockets for continued investment. Perhaps Europe's industrial policy makers should imbibe from Airbus's historic example of success less often and instead take a more sober look at the numerous European industrial policy projects that failed in the digital domain—at considerable expense.

As a major industrial policy project, GAIA-X seems policy-making through symbolism, and a very expensive variety at that. Europe's economic and digital policy makers want to show ambition. They want to demonstrate "digital sovereignty." Governmental decision makers believe they know which technological solution will move Europe forward, and how it can take flight under state direction. European taxpayers are highly unlikely to see an IT Airbus take off, let alone the launch of a digital moonshot. Most likely, the retrospective question about GAIA-X will also be this: Is this a case study of misconceiving the root of the problem—of using a hammer to tighten a screw? Shouldn't it have been the state's responsibility to create a better environment for data-based innovation without giving preference to any particular technology?

Darwin's Sea

In his late work, Lewis Branscomb, former head of the US National Bureau of Standards, chief scientist at IBM, and professor at Harvard University, introduced an interesting concept into research on innovation: the Darwinian Sea. Branscomb referred to the waters full of challenges and opportunities that innovative entrepreneurs enter once they've shown they're working on an idea with great potential, but when there's still a long road ahead in the Schumpeterian sense to succeed in bringing the idea to market. While other researchers focused on imagery of risk and failure during the innovation stage, Branscomb consciously chose a metaphor from evolution. The Darwinian Sea teems with a wide variety of species. There's a huge diversity of colorful, hungry fish frolicking about in it. There's also plenty of food. Sometimes animals can cooperate to achieve their individual goals. But competition is also fierce and, according to Branscomb, necessarily so. Innovation is a competition in which better ideas with better implementation prevail over other ideas.

Branscomb chose the image of the Darwinian Sea precisely because it's a powerful reminder that a healthy environment for innovation isn't a monoculture—or an environment without challengers and challenges. The wide diversity of different life forms, all adapted to survive and thrive, only to be challenged again by change, is the core element of the Darwinian Sea. Only then will new industries emerge, which in turn will prove rich in innovativeness.

For decades, the Silicon Valley has been a Darwinian Sea. It has fostered and facilitated the birth of a huge variety of different technologies and business models, as well as many different modes of cooperation, coexistence, and competition. Much like a rich ocean environment, the Darwinian Sea has created countless winners, as well as losers. For decades size alone wasn't a good predictor of success. Much like in the actual ocean, small prey can become quite successful in surviving, even challenging large fish. And the entire system is nourished by a constant flow of information and ideas, much like plankton nourishes larger ocean life.

The Darwinian Sea is anathema to traditional industrial policies of subsidies and protectionism, of national champions and deep public pockets that support and sustain them. The Washington bureaucracy did not build Silicon Valley, and national protectionism did not sustain it. Silicon Valley holds a lesson for policy makers around the world, intent to spend (with taxpayers' dollars) their way to High Tech success. But the enemies of the Darwinian Sea aren't just proponents of outdated industrial policy-making. For around ten years now, Big Tech in the United States has increasingly been draining the Darwinian Sea. Similar developments have taken place in China. In both instances, Big Tech companies are charging more than they would in a truly competitive environment while no longer being as creative as they claim to be. Instead, they're slowing the pace of innovation.

As nations and societies around the world are looking at how to facilitate and maintain High Tech innovation, neither an old-fashioned industrial policy approach nor the current Silicon Valley model of Big Tech oligopolies offers a sustainable path forward. Rather than invigorate innovative activity, the former will turn a few incumbent high-tech companies into "champions" dependent on receiving public funds and overcharging their customers while the latter will lead to a similar anticompetitive ecosystem that favors monocultures and stifles innovation, albeit without wasting public funds.

A far better answer can be found by reverting back to the image of the Darwinian Sea: a rich and diverse ecosystem that encourages experimentation and adaptation. And by returning to Schumpeter.

The Darwinian Sea is ideal when the trajectory of technological development is unclear and disruptive, radical innovation is anticipated. In such circumstances, the solutions are not obvious, and incumbents, focused on incremental change, may be unable to identify them. Facilitating and encouraging many and widely different ideas is key. This gives an advantage to organizations and localities featuring individuals possessing a wide variety of backgrounds, skills, and views. This isn't something that Silicon Valley has a monopoly on. Nor is it something that only predominantly large and established organizations offer. In fact, it is quite the contrary, as established incumbents tend to create cultures of conformity rather than of diversity, of efficiency rather than true radical innovation.

In contrast to the various stages of the industrial revolution, in digital times most great ideas no longer necessitate enormous capital investment to innovate. With cloud computing at commodity prices, an Internet with a global reach, and a sea of smart computing devices in the hands of billions of consumers, it's no coincidence that Reid Hoffman, co-founder of PayPal and LinkedIn, and Silicon Valley investor, titled his recent book *Blitzscaling*, highlighting the possibility for digital companies to scale with speed, not weight. But capital and commodity technology are only a necessary, not a sufficient, condition for success.

The Darwinian Sea is rich in plankton. This vital resource for the ecosystem is so ubiquitous that fish only need to open their mouths and wait for it to be delivered. It's as plentiful as water—or air is to us. It's the wide availability of the most crucial resources that drives life to the Darwinian Sea, that enables experimentation, and leads to innovation.

The Darwinian Sea thrives when societies welcome innovation, even if this entails a bit more risk. Schumpeter highlighted the importance of entrepreneurs. He was right, but only partially. In addition to individuals with new ideas and a willingness to try, the societies around them need to be ready for and supportive of them. Innovation entails risks; it also leads to countless experiments that will fail. This risk doesn't go away, but communities and societies can redistribute parts of it—for example through investments in supportive infrastructure such as transportation, housing, and education, or through adjustments in the legal system that shifts liability and cost away from innovative entrepreneurs.

This is hugely encouraging for regions and nations around the world. Branscomb's and Schumpeter's message is that no place has a monopoly on the future, and the remnant structures of the industrial age are no suitable predictor for success in digital times. Even past triumph in the digital realm does not imply continued success, as evidenced by the slowing of innovation and business dynamic in the Valley.

But at the same time, this suggests a significant challenge. There is no simple, easy to enact blueprint for sustained innovation. To do this on a systematic basis requires a comprehensive policy framework at multiple levels. It needs investments in the right infrastructure (think minds not streets), the right regulatory framework (think of managing risks rather than avoiding them), and, most importantly, in the spirit of Schumpeter and his concern of entrepreneurs deprived of resources essential for success, it necessitates an innovative policy solution: opening up access to the key resource of innovation today, data (think of it as the plankton of the data age). That is the focus of the next two chapters.

5. *Might and Machines*

In the fall of 2019, Facebook's leadership surely must have hoped to revert to their core mission: attracting even more people onto their platforms and earning even more money out of their users through personalized advertising. The months before had been quite turbulent. Mark Zuckerberg had been summoned to appear before Congress and the European Parliament. The US Federal Trade Commission had imposed a $5 billion fine on the company for disclosing user data to infamous British data analytics company Cambridge Analytica, making false statements about the use of facial recognition, and related problems. Facebook had promised to clean up the mess by shaking up internal processes and its organizational structure, and it had agreed to submit to an external audit. Only then did the regulatory tidal wave battering against arguably the least popular of the Big Tech companies appear to abate.

But just at that moment the Democrat primaries were heating up. Progressive candidates, above all Senator Elizabeth Warren, an ex-Harvard professor well-versed in digital issues, raised questions about the power of information in dramatic terms. In March 2019, Warren had already announced that as president she would break up Big Tech companies. Amazon would no longer be allowed to be both a shop and a marketplace and would be forced to divest organic grocery chain Whole Foods. Google would have to part company with Waze. For Facebook, left-leaning Democrats determined that its acquisitions of WhatsApp and Instagram would have to be undone—a potentially enormous technical and organizational challenge for the company.

Such calls for corporate breakups, the most radical tool of antitrust law, were not entirely new to Facebook lobbyists. In his book *The Curse of Bigness: How Corporate Giants Came to Rule the World,* Tim Wu provided the historical blueprints by retracing famous earlier breakups of Standard Oil and AT&T. But just as the worst swell of public outrage over Facebook subsided, the online technology magazine *The Verge* published a recording of internal discussions between Zuckerberg and other Facebook employees. Speaking of the Democrats' regulatory proposals, Zuckerberg commented that if Warren was elected president, it would "suck" for Facebook, because they'd probably get sued by the US government. But, according to Zuckerberg, antitrust action wouldn't be that bad. He'd wager a bet that Facebook would win.

Facebook had once again become the perfect target for its opponents. The recording reflected precisely the kind of arrogance that had given tech titans, once celebrated as pioneers of the information revolution, a reputation on par with that of Wall Street bankers. In retrospect, however, it's clear that Mark Zuckerberg was even more prescient than he imagined when addressing his employees.

Recent US presidents and their administrations had not been particularly tough on Big Tech. This may be a consequence of the fact that Silicon Valley has donated heavily to presidential campaigns. But in part, recent administrations also had other priorities.

George W. Bush had two wars to deal with and little appetite to tackle a contentious political issue, especially since his party saw Washington's role more in supporting large corporations than enforcing regulation to foster competition. President Barack Obama had a different "distraction" to deal with—a global financial crisis. When battling the prospects of a long economic depression, pursuing regulatory action against the few companies that do comparatively well and create jobs was fairly low on the priority list even of fairly liberal Democrats. Facebook's major acquisitions, from WhatsApp to Instagram, may have raised some eyebrows and prompted an occasional Congressional hearing, but did not lead to sustained regulatory headwinds.

The rhetoric changed with the ascendence of Donald Trump. The president himself took to his favorite weapon, Twitter, to rail against Big Tech and what he perceived as these platforms' liberal bias. His administration pursued Google and Facebook for antitrust violations. Yet, this wasn't a watershed moment for regulatory Big Tech backlash. As in other areas, the Trump administration lacked persistence and effectiveness to turn rhetoric into regulatory outcomes.

There may have been, however, another more strategic reason for Washington's reluctance to rein in Big Tech: these superstar companies project American global tech dominance. And that is ultimately good for the US economy and its geopolitical power. Declining competition seemed to be not only a comparatively small price to pay, but the necessary prerequisite for American digital platforms to succeed throughout the world according to the economic logic of scale and network effects—even though some voices already questioned whether this game was sustainable in the long term.

When Joe Biden won the 2020 US election, many in Big Tech must have sighed in relief, not only because of their own political leanings, but because the incoming president, unlike Senator Warren, had not been an antitrust firebrand during his long political career in Congress and as vice president. Biden was perceived as a centrist rather than an activist liberal, and one that wouldn't rock the boat unnecessarily. In Silicon Valley, the hope has been that Biden could repeat what George W. Bush's administration did two decades earlier when it effectively dropped its antitrust fight against Microsoft in the courts.

And they were not alone. Weighing the pros and cons of decisive regulatory action, some American antitrust lawyers made the case against comprehensive regulatory action. They advocated in favor of accepting the acquisitions of the past and abandoning a likely hopeless struggle to reverse them. In their view, it would be better to focus on the future by enacting a series of far more incremental measures to ensure entrepreneurs no longer prefer selling out to Big Tech companies. For instance, they suggested IPOs should become significantly easier processes for start-ups, and IPO profits for entrepreneurs and venture capitalists should be taxed more favorably than when selling out to a large competitor. They also called for promoting "venture debt," or bank loans for start-ups, that don't require entrepreneurs to cede ownership, unlike venture capital. That way, they won't be easy prey for VCs pushing them to sell out quickly.

Another proposed measure was a "post-acquisition lockup." Former equity holders would only receive the proceeds of their divestment when a start-up's technology or service reaches certain mutually agreed milestones within the acquiring corporation, along the lines of "earn-out agreements." This incentivizes the (large) buyer to create and maintain the conditions for the (smaller) company being acquired to be successful within the new corporate context. It's an interesting idea, because it might largely put a stop to "killer acquisitions" or anticompetitive exits that involve buying innovative companies with the intention of shutting them down.

But can a package of relatively minor measures focused on M&A really solve what is fundamentally a structural problem? To the shock and surprise of some antitrust lawyers and Big Tech alike, President Biden did not think so. And he wasn't alone.

Break 'em up?

Early into his term, President Biden appointed Tim Wu, who had argued in favor of breaking up Facebook and written popular books on the dangers of Big Tech market concentration, to the National Economic Council as a special assistant to the president for technology and competition policy. Putting one of the most outspoken advocates of Big Tech trustbusting into a top advisory role is a powerful signal the Biden administration is taking a far more confrontational course.

Wu isn't alone. His appointment was followed by the choice of Lina Khan for chair of the Federal Trade Commission (FTC). Khan's youth—she was in her early 30s when nominated—belies her intellectual power and political credentials. A professor at Columbia Law School like Wu, Khan had authored influential papers on the need to fight Big Tech's unchecked power. And she had explained why existing antitrust law was ill equipped to deal with Silicon Valley platform providers. But Khan isn't just a Big Tech critic; she also offered a radical solution: regulate Big Tech companies as utilities, much like electricity providers or the venerable AT&T before telecom deregulation. With Khan at the FTC and Wu as advisor having the ear of the president, Big Tech could be in serious trouble.

Not just antitrust experts serving in government like Tim Wu and Lina Khan fear that the monopolistic structure of American tech dominance could turn into its Achilles heel. Think tanks and advocacy groups on both left and right have been joining the critics. Disruptive entrepreneurs and venture capitalists such as Elon Musk and Peter Thiel regard the well-rehearsed dance of Big Tech and venture capital with increasing skepticism, concerned that the intricate choreography is thwarting the next generation of disruptive founders and technologies. Taken together these voices are calling on and supporting regulators and legislators to prevent the most obvious cases of large companies removing potential competitors from the market by acquiring them—cases comparable to Facebook's takeover of Instagram or Google's acquisition of Waze. And they call on venture capitalists to take on the role for which Joseph Schumpeter originally conceived this class of investment capital, the role that the venture capitalists on Sand Hill Road in Menlo Park fulfilled up to the first decade of this century: financially support the bringing to market of new, radically better ideas and then enable them to be scaled up.

The antitrust tide is rising in the United States. And yet it's questionable that well-intentioned activist regulators bolstered by broad public support will succeed. The challenge is a combination of the structural and the political. As Lina Khan herself argued, existing antitrust laws are less than useful. Big Tech may not have violated them sufficiently to warrant breaking them up. And other powerful measures, such as declaring them utilities, require legislative action. Given the delicate power balance in Congress and hyper-partisan politics, it's likely that such bold legislative proposals would not get enough votes to become enacted. The political factions may agree on the problem, but they are far apart on the solution. The left wants an effective remedy, while the right insists on the importance of market forces and worries about antitrust action micromanaging economic activity. That leaves a fairly narrow corridor of acceptable incremental legislative steps, such as "post-acquisition lockups." This may be politically palatable, but insufficient to achieve real and sustained success.

The truth is that the current game based on exit strategies works only too well for everyone involved, at least in the short term. The monopolists continue to increase their rents. Entrepreneurs get rich quickly. Venture capitalists reduce risk by optimizing their investments for exiting through a sale. And government? It too earns money on every "Goliath buying David" transaction. Preventing such transactions causes annoyance for everyone involved. Any politician mounting a serious attack on Big Tech USA exposes themselves to the charge of endangering the great successes of American technology companies on global markets—a charge few politicians could fend off.

Despite renewed resolve by the Biden administration to get serious against Big Tech overreach, substantial change still seems elusive in the United States. In contrast, European antitrust authorities have been far more active. The billion-dollar fines lobbed at US Big Tech by Commissioner Vestager's team surely sound impressive. But, as we mentioned, most of them were reduced on appeal to an amount that the superstar companies with huge cash reserves and skyrocketing profits could easily afford. The European Parliament may not suffer from hyper-partisanship and be willing to strengthen antitrust rules, but their effectiveness is limited by the very fact that almost all Big Tech is not European. At best, Europeans might prevent US Big Tech from buying up innovative European start-ups; the necessary laws for this are increasingly being enacted. But that will do little to break Big Tech's information power.

The challenge faced by European regulators is shared by regulators around the globe, from the Asian Tigers to the Global South: how can national regulators effectively counter the information might amassed by Silicon Valley superstars? Sure, one could prohibit US Big Tech from operating. But that would deprive the local economy of valuable services. For most nations, such binary disengagement is not an option. And for nations that to an extent can and have disengaged, such as China, their homegrown Big Tech companies confront them with similar problems. The huge fines levied on Alibaba in 2021 surely are surprising for outside observers, but they, too, are targeting symptoms, not the root cause of Big Tech's power.

Sooner or later, regulators and legislators will have to confront the real problem of reining in Big Tech: whether we look at Draconian measures like breakups or incremental ones like fines and acquisition lockups, these target the symptoms of Big Tech's information power, but do little to undo the structural advantages the digital superstars possess. It's little more than cutting a head off Hydra, only to see a new one grow.

To tackle the structural advantage, we have to remember Schumpeter. Schumpeter's nightmare was that the capacity for innovation would become concentrated within a few large companies. This would lead to a downward spiral of innovation, as major players have less incentive to be disruptive and far more reason to enjoy market power. Contrary to Schumpeter's fear, this concentration process didn't occur after World War II, mainly because entrepreneurs had access to abundant capital and could thrive on disruptive ideas. They stood a real chance against the large incumbents of their time, a role more than a few of them took on themselves. But money is no longer the scarce resource limiting innovation. What's scarce today is access to data. More precisely, such a scarcity is being artificially created.

In the data economy, we're observing a concentration dynamic driven by narrowing access to the key resource for innovation and accelerated by AI. The dynamic therefore turns on access to data as a raw material. Economic policy to counteract market concentration and a weakening of competition must focus on this structural lever.

If we want to avert Schumpeter's nightmare, preserve the competitiveness of our economy, and strengthen its capacity for innovation, we have to drastically widen access to data—for entrepreneurs and start-ups and for all players who can't translate their ideas into innovations without data access. Today, they can only hope to enter the kill zone and be bought up by one of the digital giants. If data flows more freely through broader access, the incentive to use data and gain innovative insights from it increases. We'd turbocharge our economy's capacity for innovation in a way not seen since the first wave of Internet companies. We would also learn more about the world, make better decisions, and distribute data dividends more broadly.

Data for All

Among the numerous remedies—from incremental to radical—that have been proposed recently in the United States to reign in

Big Tech power, drastically broadening access to data hasn't yet played a prominent role. This is surprising given that access to data is a pro-market and pro-competition strategy, whose primary aim is not to cripple Big Tech but to enable digital startups and grown-ups alike, facilitate competition, and broaden innovation. It would align well with the fundamental principles of free enterprise and open information flows that have inspired the republic since the days of Benjamin Franklin.

Even more surprising is how broader access to data has intrigued European policy makers. One might assume that the importance afforded in Europe to data protection may constrain the appetite for access legislation. But in February 2020, the EU Commission issued its digital and data strategy, and in an op-ed accompanying the strategy, EU Commission president Ursula von der Leyen stated, "We want that data to be available to all—whether public or private, big or small, start-up or giant," and, she added, "as part of this, big commercial digital players must accept their responsibility, including by letting Europeans access the data they collect. Europe's digital transition is not about the profits of the few but the insights and opportunities of the many. This may also require legislation where appropriate." The head of the EU Commission could not be clearer about her strategic aim of broadening access to data, where necessary through new regulatory mandates. Of course, the European Union runs on political consensus, so the Commission's ambitions are not always realized, at least not in the first round. But von der Leyen had clearly put down the gauntlet.

The EU Commission's stance makes sense, though, when one looks at the economic situation. The European hope that so-called unicorns—startups valued at more than $1 billion—would create serious competition for established digital monopolists hasn't been fulfilled. Over the last ten years, almost none of the major platform markets has seen start-ups successfully shake up the status quo with new ideas. It may be gratifying that at least in some niches, European digital platforms such as Spotify, Trivago, Wise, or Zalando have gained some traction, often applying business models copied or slightly adapted from American trendsetters. Some of them are even successful outside of their home turfs. But where a true leap forward in innovation was beckoning, such as with AI leader DeepMind in London, US superstars simply reached into their financial war chests and bought the start-ups off the market. In the case of DeepMind, Google pounced in 2014. Ever since, "European" DeepMind technology has not only made possible Google's spectacular success against the world's best Go players with its AlphaGo engine but is now also powering various aspects of Google's Now AI assistant.

Meanwhile, in a growing number of more traditional industrial sectors, young and data-rich companies from Silicon Valley are causing increasing difficulties for major incumbents long on tradition and short on data expertise. Examples can be found in abundance in banking and machine building, in retail and tourism, in mobile communications as well as medical and health. To arrest and reverse this shift, the EU Commission seems to have concluded that the continent must reinvent itself economically, socially, and politically. More use of and broader access to data, the most important raw material for innovation in the twenty-first century, seems to be a key element.

But Europe's position isn't unique. Societies and economies in Asia, Africa, and Latin America are equally reliant on US Big Tech, and, to a smaller extent, Chinese equivalents. Even in the United

States, Big Tech is concentrated in the Valley and up and down the West Coast. Most regions aren't reaping the digital dividend that Big Tech accrues, but their inhabitants pay the monopoly rents to the digital superstars like everyone else. As we explained, innovation activity and business dynamism are sagging in the Valley, but the situation isn't better in the Rust Belt, the Sun Belt, the Midwest, or the Deep South. Digital markets are no longer competitive, and the technology sector is no longer as innovative as it has been. Arresting this trend and rekindling the innovation flame is essential for the United States, as much as elsewhere. This is a global problem.

Choosing the appropriate policy remedy will define not only how economies fare, but also how societies value innovation and view market competition. Is this about opening opportunities or punishing digital incumbents? Are we projecting into the future, or re-prosecuting the past? Admittedly, wider access to data won't appeal to those who now get to corral data for themselves. The gloves are already off in the no-holds-barred battle of the lobbyists.

Advocates for Big Tech sometimes claim that mandated access to data amounts to expropriation. This is groundless propaganda. Data cannot legally be owned like physical property, and for compelling economic reasons. Data is an informational good, not a physical one like a car or a house. Physical goods can typically be used by just one person at a time. If I'm sitting in a chair, no one else can sit there. For data, however, it's different. Data can be used by many people at the same time without necessarily reducing the value for any one of them. For fans at a concert or viewers of a movie, after all, the value of the music or the film doesn't decrease if there are more or fewer people in the audience. A novel doesn't lose its drama if many people read it. Economists call it non-rivalrous. If you broaden access to data, little if anything is taken away from anybody.

Of course, having exclusive use of data leads to competitive advantages. Information asymmetries are one basis of value creation in the information economy. But what's often overlooked is that not every user of data will use it for the same purpose and in exactly the same way. Hence, many uses of the same data are not directly competing with each other. When the same data is used for different purposes, economic value accrues that otherwise would not be created. Reusing data is not only economically efficient, it also conserves resources and is therefore more sustainable than using collected data only once or not at all. In addition, it's often unclear whether a particular data use will bring insights. Precisely because we don't know what will add value, it makes sense to utilize data for multiple purposes.

But even if data is used for the same purpose, the economic advantages outweigh any drawbacks. It leads to competition around how best to use data for a particular end. The winner is not whoever secures exclusive access to raw material and hopes to establish a monopoly on it, like the owner of a diamond mine trying to control supply and drive up prices. Instead, the winning company is the one that makes best use of the data. It's the one that best harvests potential insights contained in the data. Access to data ensures that economic actors compete less on controlling raw materials, and our economy thus avoids repeating the history of control of mineral resources such as coal and oil. Instead, competition will be focused on *use,* on the analytics stage, and thus precisely on what so far has been lacking most of all in many companies. Some companies afraid of entering the data economy may object. They may fear they'll have to make their data accessible to others who can derive greater value from it than they can. It's the old envy of the raw material suppliers over the value added in finishing, which leads to absurd attempts to prevent it and in turn creates inefficient economic deadweight.

Broader access to data redistributes the power of the few by empowering many to derive value from the data. It promotes competition and strengthens the economy, as well as nonincumbent competitors. These companies, so prevalent in many economies, can only be successful in the long term if they use data. To do this, they not only need expertise, but also have to understand that using data, rather than simply collecting it, adds value. This awareness is still lacking today in most economies. Mandated access to data creates clear economic incentives to develop the right mindset and use data. Thus, broader data access will hasten the sorely needed transformation of digital laggards. And when sufficient data is available, there is no excuse for further delaying a firm's digital transformation.

Data Market Failure

Sceptics of broader access to data often contend that this would reduce the incentive to collect data, leading to an era of data scarcity. That's nonsense. First, we collect around five times more data today than gets used even once. Another way of putting it is to say over 80 percent of data collected worldwide doesn't get used a single time, does not create value, but in fact destroys value by incurring collection and storage costs that are not offset by additional insights. If we are not reaping any benefits from data, yet still collect it, the argument that collection requires an economic incentive seems fundamentally flawed, to put it mildly.

In fact, reality suggests collecting data hardly requires any substantial economic incentive. It's easy to understand why: Data is often collected incidentally when and where it is being produced, and at low cost. Wider access to data won't change that. On the contrary, the potential value generated by third parties using data should only increase the incentive for data collectors to not just gather but also utilize the data they amass. This will intensify data use in the economy—which is precisely what's urgently needed, considering the low usage rate overall.

Data reflects only certain aspects of reality, depending on how it was collected. This means that each data set, even when covering the same phenomenon, is always going to be a little different from any other, as it's collected using different sensors or taken from different perspectives. On the one hand, this means that every data set by itself is quite incomplete so it contains errors, which can lead to false conclusions or overgeneralizations. But if you broaden access to data to cover many different sources, the results may improve as systematic errors in individual sets could cancel each other out. This makes open access especially rewarding for small to medium-sized companies: it gives them access to diverse data sets and thus improves the insights gained from them.

Some digital policy makers and trade bodies argue that a market for data would be preferable to broader access because collectors could recoup the cost of gathering data, and the market would signal through price what data should be collected. On the surface, this seems a valid argument. Who could desire increased innovation and competition yet by the same token oppose use of the market as an efficient transaction and coordination mechanism? But the argument is nevertheless flawed, for several reasons.

The market is indeed an effective mechanism for allocating scarce resources. That's only true, however, if there is sufficient information about supply and demand flowing through the market. But this is a structural problem with data. With data it's unclear at the outset whether the insights one hopes to elicit from it are in fact contained within a given data set. The purchasers of a data set (or more precisely: of access to a data set) might thus be forced to buy a pig in a poke. If there's not enough information about the nature of the product, the market can't establish a correct price. Consequently, buyers would pay either too little or too much, and, like a lottery, sellers either hit the jackpot or don't earn a penny. Economists would call such a market inefficient, because it doesn't adequately perform its role as an allocation mechanism. And the people affected would say: I didn't pay for this, I want my money back. In such a market, buyers and sellers alike lose confidence and no longer utilize it. Then, instead of no gain, there's a lot lost.

The challenge posed by insufficient information is also captured in what economists call the "lemons" problem. If buyers do not have comprehensive information, they have to assume that what is on sale is inferior due to some hidden shortcomings and offer a lower price. This prompts sellers with high quality products to exit the market because they won't get what they think is fair value for their goods, thereby only increasing the ratio of inferior products. Used car markets are like that, because a car's faults are not obvious to buyers. But data markets would likely suffer from similar problems, leading potentially not only to lower prices but less liquidity and thus overall allocation inefficiency.

To make matters worse, markets can only function as efficient allocation mechanisms if there's an exclusion right to data—for example, a right of ownership. Otherwise, one can't stop third parties from simply grabbing data others have collected. But currently there is no such exclusion right to data. It would first have to be established to make data markets possible. The legal details of such an exclusion right would be challenging, because, among other things, an exclusion right to data isn't obvious to others, unlike with many material goods. Above all, however, it's unclear how such an exclusion right should be designed. If it closely resembles property, then it's fairly clear what the "buyer" is getting, but, by the same token, the use of the data would be difficult to restrict to specific purposes or specific users. But this is exactly what many data providers want in order to prevent their data from being passed on to their direct competitors. If a right to own data doesn't make these restrictions possible, then many data sets won't be offered for sale on a data market in the first place.

On the other hand, the less an exclusion right resembles fully fledged ownership, the easier it is to restrict use and disclosure. This could lead to data holders being more willing to make their data sets available. But it also entails tougher negotiations with potential buyers about exactly how they're allowed to use or disclose the data to third parties. It increases transaction costs. Nobel Prize winner Ronald Coase showed in the 1960s that when transaction costs are high markets can't fulfil their allocation role effectively. This is why all large-scale attempts to establish data markets have failed so far. There was either insufficient supply or too little demand, or completing the transaction was too complicated and costly. Even Microsoft had to accept this reality and abandon its vision of a large data market as part of its Azure cloud platform in 2017, only seven years after its introduction.

Markets help allocate *scarce* resources. But it's not data that's scarce—it's its use. If one had to purchase data, the use of data would become more expensive. That's exactly the opposite of what many economic sectors need right now. Remember, about five times as much data is collected than gets used even once. As the incentive to use decreases, competition and innovation suffer. At the same time, data users would raise prices for their services, since they have to purchase access to data from its collectors. Ultimately, customers would foot the bill in the form of monopoly rents paid to the very companies already hoovering up most of the data around the world. That would only create a new version of Schumpeter's nightmare.

Because a data market can't fix the problem, because we need incentives to use rather than collect data, and because broader access to data leads not to expropriation but to empowerment, the compulsory opening of data sets is the logical response to Schumpeter's nightmare. Economies that are data-rich thanks to broad access could turn into the world's most innovative places. Regions with open access to data could create a Darwinian Sea (as envisioned by Lewis Branscomb) for digital innovation, rich in both nutrients and competition, as only this combination will beget more novel products and ideas and help these regions to regenerate through innovation.

European education experts and human resource managers lament that while universities and research institutions are training young data scientists and experts in machine learning, by the time they've earned their doctoral degrees, they've leapfrogged over to the American superstar companies and their hefty pay checks. This analysis is spot-on. The engineering departments of Big Tech companies are full of young people with European passports who received their basic training in Barcelona and Paris, Budapest and Sofia, London and Munich. That, too, isn't a uniquely European problem. Quite the contrary: Big Tech relies on a steady stream of young data analysts and coders from China and India and many other nations, who had come to the United States for their terminal degrees and then were hired by the superstars in the Valley. The salaries Big Tech pays are surely enticing.

Even though the Valley remains the primary magnet for such talent, it is no longer the only destination. A significant portion of Chinese students returns to China to join Chinese Big Tech, and increasing numbers of talent from Europe and Asia (and even America) are joining them. This may be the right choice for them, but these talents are missing in the nations they left.

This dynamic isn't set in stone. It can change. Talents are looking for opportunities, and opportunities beckon where innovation thrives. Regions awash with data thanks to broad access could offer an overall package more attractive to talented minds. For this to happen, however, a region's data policy would not only have to override all the doubts of those in business who are trapped in the past and see access to information more as a danger than an opportunity. It also would require looking at privacy regulations from a different perspective. This may be doable for nations in America, Asia, and Africa. It's possible for Australia. But for Europe, it would be quite challenging, even though it would mean returning data protection to its roots.

The Religion of Data Protection

The world's first data protection laws, in the German state of Hesse and in Sweden, had two goals: to protect individuals from harm caused by a misuse of their personal data, but, at the same time, to balance informational power by stipulating access to data. The Hessian data protection law for instance gave parliament access to data from the state's executive branch. And in Sweden, free access to information had a long tradition as an antidote to the concentration of informational power, predating data protection by centuries. As a result, information on income and assets of citizens, for example, has long been open and accessible so social inequalities cannot remain covered up.

The goal of the founders of data protection regulation wasn't to prohibit the use of data but, rather, to enable use as long it did not harm individuals. It's precisely the legal boundary between meaningful and harmful use that instills confidence in citizens in the use of their data. Data protection was seen as furthering multiple ends: to protect individual freedom, enable data-based decisions (especially business decisions), and promote political participation and democratic governance. The protection of privacy and the dismantling of information asymmetries are therefore not a contradiction in terms. Data protection and data access are two sides of the same coin.

In Scandinavia, this duality was recognized in law and in practice. It's no coincidence that Stockholm emerged as the most successful start-up metropolis per capita in Europe or that Scandinavian entrepreneurs turned Spotify, Skype, and the games empire Mojang (including Minecraft) into some of the few world-class European digital platforms (two of which were bought by Microsoft). In contrast, national legislators in the rest of Europe reduced data protection to limit the use of personal data. In theory, this should not have resulted in regulation slowing down digital innovation using nonpersonal data. But, through culturally anchored practice and economic impact, that's exactly how it turned out.

"If the EU had an official religion, it would be privacy," *The Economist* put it in April 2020: "A devout priesthood of EU officials and politicians preach that only their privacy laws can lead to salvation. Holy texts, such as the GDPR or the ePrivacy Directive, are held up as wisdom the whole world would be better off following." For European citizens, the text continues, the sermons from preachers of this religion are often "about as intelligible as a mass in Latin."

In discussions about broadening access to data in Europe, we too have experienced that even faintly questioning the dogmas of data protection can mean being branded a heretic. But, by the same token, more and more people are no longer able or willing to listen to the moralizing litany of the high priests of data protection. At home, in public life, and at work, they're discovering that with data protection, too, political dogmas can do more harm than good. This became brutally apparent during the coronavirus crisis. The crisis raised questions that made the benefits of a wealth of information for individuals and societies clear for the bulk of Europeans not versed in data-speak.

For instance, is it more important for children in lockdown to be taught as a group, at least by video conference? Or is it okay if a parent can prevent the whole class from learning together on the grounds of protecting the privacy of their child? Is it right for data protection officers to threaten to fine teachers for providing digital tools for home schooling? Is it preferable to prohibit employees from working from home during the pandemic and instead require them to return to a virus-contaminated open-plan office because the company's time recording software isn't allowed to operate over the Internet, allegedly for data protection reasons? And does it make sense to create a contact tracing app but deprive it by design of collecting useful data? The COVID19 pandemic brutally exposed Europe's digital weaknesses when it comes to the availability of these systems and expertise in using them. Most importantly, it became abundantly clear that the principle of minimizing data use comes at a high price: dysfunction. In an age of a wealth of data, it simply makes no sense to impose structural scarcity on the most valuable and potentially most useful asset.

Since May 25, 2018, all EU member states are obliged to enforce GDPR. But European data protection's original aim of empowering individuals, organizations, and institutions through a right to information access has been almost completely eradicated. Almost nowhere among the nearly one hundred articles in the GDPR are any provisions that tackle information imbalances society-wide and counteract concentrations of informational power. A last, humble remnant of this tradition survived in the data portability right—that is, the right of consumers to request processors of their data to transfer that data to another provider. But it's almost never used in practice.

In its essence, GDPR is an act of legal disempowerment. It regulates what data may not be stored, processed, or analyzed, and by whom. And it exaggerates the importance of obtaining consent. In effect, a handful of people can block potential benefits for the many. In certain contexts, this may be appropriate and necessary. But in practice the consequences are much broader. For most nonpersonal data, some connection to personally identifiable information of some kind can be made, even if legally irrelevant. And where doubt exists, the fear that a data protection officer might see things that way is often sufficient for organizations and companies not to utilize the data they have. Not only in pandemic times, the resulting deliberate and artificial scarcity of data may turn out to be economic and societal madness.

Compared with global digital reality, Europe's quasi-religious variety of data protection has developed a bizarre momentum of its own. The superstar companies have much more data than European companies. The few data-driven European companies have significantly more data than the digital stragglers. In economic terms, it would be much smarter to *empower* those who lack data to use it and thus to develop expertise in its use rather than stopping them in their tracks while giving carte blanche to digital monopolists to continue milking their data troves. Such a paradigm shift in data policy, from artificial scarcity and disempowerment to broad empowerment, would reduce today's concentration of informational power and raise the great game of innovation to the next level.

The case of data protection in Europe offers an important warning. Nations and societies that have narrowed their policy focus too much, lose the bigger picture. They only see the trees, no longer the forest. They cherish data protection and data scarcity as a value in itself, rather than what it originally had been—a means to ensure that gross information imbalances would not develop in markets and society.

This is dawning on policy makers. It's probably why the EU Commission is steering so starkly in the direction of data use and data access. And it is a reminder to other nations, from Asia to the United States, that our societal realities aren't a given. They can be and are shaped by policies—and those that choose well, win.

Data Innovation

This is where we have got to: Too many data resources aren't being mined. Those who extract and store data often have no use for it. Such data therefore creates no value for either economy or society. But this data still exists because it accrues incidentally, or gathering it is so inexpensive that it gets stored even without a specific use in mind. The fear that companies, if they were legally obliged to permit access to some of their data, would no longer have an incentive to gather data has been greatly exaggerated. And if information-based domination derived from exclusive access to data disappeared, economic incentives for companies to invest more heavily in data expertise, data-based innovation, and data-driven business models increase. Platform companies would no longer be able to monopolize data. Instead, in the tradition of Schumpeterian innovation, competitive advantages for companies would arise from the information economy's most important skill—the ability to extract insights from data to create better products and services that benefit customers and society.

A comprehensive mandate for broad data access empowers individuals, companies, and institutions to utilize this resource. While exclusive access to data by today's data monopolists could be broken, they would also gain access to more data themselves. In background discussions, some representatives of traditional industries tell us that, in principle, a more open access to data is a great idea as long as it's designed so that they can access Big Tech data, but not the other way around. But that's like wanting to swim without getting wet. Such one-sided data access would not only be unfair. It would also be hostile to innovation as it would leave large troves of data unused. It would contradict values of openness and international cooperation. And it may precipitate new trade wars. Data mercantilism is a bad idea. It conceives of "sharing" data with others as primarily a zero-sum game in which mutual access signals loss of control and leads to competitive pressures that must be prevented at all costs. This is wrong. Not only is it destructive, but it also shows a shocking lack of self-confidence.

If everyone shares data, everyone wins. Economically speaking, data is a lot like air. It's all around us, and at the same time we need it for economic prosperity, for innovation to make better decisions. That's why access to it should be as natural as access to the air we breathe. It may sound like an airy-fairy vision. But it's feasible and can be enacted. With an attitude of facilitating data use in business, politics, and society, digitalization will finally be able to fulfil one of its grand promises. The information engines of the few will become instruments of empowerment for all.

6. *Access Rules*

In the late 1950s, the Santa Clara Valley in northern California was known for its fertile soils, not for microelectronics, semiconductors, or computer pioneers. In 1951, an "industrial park" had already been established at Stanford University. But the transition from an area dominated by orchards, fruit trees, and a renowned university to Silicon Valley got off to a faltering start. Until the mid-1960s, the American epicenter of high-tech was on the East Coast, on the outskirts of Boston. It was known as Route 128, named for a Massachusetts state highway. Dozens of tech companies sprung up along this "American technology highway" under the (figurative) walls of Harvard and MIT, funded by venture capital from Georges Doriot and others. Digital Equipment Corporation (DEC), the company he invested in so successfully, was just one of them. Data General, a minicomputer pioneer founded by former DEC engineers, was another. By the end of the 1950s, more than a hundred innovative start-ups employed almost twenty thousand people. Ten years later, there were over a thousand high-tech companies, including Wang in office automation, Raytheon specializing in radar and microwave technology, and Polaroid, the instant photo pioneer. Industry observers talked of a "Massachusetts Miracle." But by the end of the 1980s, the magic had vanished.

Within a few years, most of the start-ups had gone bankrupt or been bought out. Today, DEC, Data General, Polaroid, and Wang are names found only in history books. Just Raytheon was able to stay afloat, thanks to a merger and aid from the Pentagon's deep pockets. The most ingenious thinkers at the forefront of technological progress had left the East Coast and moved west. In 1971, the journalist Don Hoefler coined the term "Silicon Valley." From that moment on, the label served not only as a pointer to the key material in the new, rapidly growing semiconductor industry of Northern California, but also to a different kind of collaboration across company boundaries.

AnnaLee Saxenian is a leading expert in uncovering the reasons for Route 128's decline and Silicon Valley's rise. Her 1994 comparison *Regional Advantage* quickly became a classic. Its analysis is amazingly succinct and illuminating, more than a quarter century later. Saxenian had long been interested in Silicon Valley. As a budding urban planner, she wrote her master thesis on the downsides of the high-tech boom in the Valley: skyrocketing land prices, unaffordable rents, insufficient living space, overwhelmed transportation infrastructures. And yet, Silicon Valley worked. What made it attractive must lie deeper, be less physical, more intellectual. Saxenian began to investigate how people in the Valley interacted, how information and knowledge was flowing. What she found struck her as remarkable—and fundamentally different to what she had experienced growing up in the Boston area. So she returned to the East Coast for a year at MIT and studied the evolution of high-tech companies around Route 128.

The East Coast tech companies were vertically integrated and hierarchically organized. They were based on the business model of industrialization and focused on economies of scale and market power. That worked well at the beginning of the digital revolution until the 1970s, but then suddenly new competitors emerged in Northern California that marched to a much different beat. These had deliberately cast aside the mindset of industrial mass production.

The West Coast's new start-up culture was based on free and open flows of information, decentralization, and competitive cooperation. The tech companies in the new technology hub, with Stanford as the elite school for talent, teamed up to secure access to the essential components of success. At the same time, they were engaging in fierce competition with each other to create value through innovation. The meteoric rise of Silicon Valley from the mid-1960s onwards was based on the cultural equivalent of broad data access: open exchange of information. The engineers from various start-ups met after work at a number of bars and cafes and shared experiences in a way that would have been unthinkable on Route 128.

In the technology they created, the young West Coast engineers relied on open standards to facilitate and perpetuate the exchange of information. Employees in the Valley often switched jobs and took their knowledge with them from one start-up to the next. The noncompete clauses in their employment contracts didn't stand a chance in Californian courts. Then there were the venture capitalists who bought their way into dozens of start-ups at the same time and actively stimulated and supported the flow of information from their seats on the board of directors. Information from Xerox PARC shaped Apple's Mac, which in turn was the inspiration for Microsoft's Windows. In the early 1990s, eight of the Valley's leading chip specialists decided to share information with each other and granted each other the right to use their patents. Building on this shared resource, they each competed intensely against each other to develop the most innovative chips. The leading hard drive manufacturers of the time, Seagate and Quantum, signed a similar pact in 1992. The aim of this cooperation on the basis of open sharing of information under conditions of tough but fair competition—later frequently referred to as "coopetition"—was value creation through innovation.

There was nothing like it on Route 128, not even a cafe where engineers would meet. No information flowed between companies; no patent pool. Company executives watched each other with suspicion. Saxenian's key point in *Regional Advantage* is that Silicon Valley's magic formula wasn't the nicer Californian climate or the proximity to Stanford and Berkeley, but the open flow of information and the decentralized, flexible organizational structures that enabled and facilitated this flow. That was the source and foundation of the dynamics of innovation that made Silicon Valley and its companies famous. Globally connecting computers via the Internet further propelled the idea of free flow of information. But as some companies turned into digital superstars dominating their markets in the first decade of this century, they broke with Silicon Valley's cultural history.

Google, Facebook, and Apple began to reap profits from data-rich platforms they had created through groundbreaking entrepreneurship and a culture of collaborative innovation. The digital giants recognized that they could rely on exclusive access to data rather than have to hope for "moonshots" (which often failed). Since then, they've increasingly resisted that culture of open information and innovation, mostly in subtle ways rather than openly. With sushi, yoga, and beach volleyball on the company campus, they seek to keep their employees on site, so they don't exchange information with colleagues from competitors over drinks after work. Open flow of information is desirable within the organization alone because on platforms with abundant data, the simple formula applies: the more exclusive the data use, the higher the profits. Mentally, Silicon Valley has moved closer to Boston's Route 128 of the 1970s. And this creates a tremendous opportunity for other regions around the world.

Saxenian went on to become the dean of the University of California at Berkeley's iSchool. In that role she succeeded information economist Hal Varian, who had become Google's chief economist. But Saxenian has mixed feelings when she considers the more recent transformation of Silicon Valley and the rising influence of the Big Tech companies. As Silicon Valley is losing some of its original mojo, Saxenian, who is sharp, perceptive, and visionary as ever, already has her eyes on other regions around the world that have taken the Silicon Valley DNA—free flows of information in a decentralized ecosystem teeming with coopetition—to heart. When asked where today the innovation engine runs in overdrive, she answers Israel, the "startup nation," where many of the original Silicon Valley qualities live on.

A Mandate for Access

As described in the previous chapter, American antitrust regulators have spent the last two decades soundly asleep, largely letting digital superstars do their own thing. However, in April 2011, they briefly woke up. The US Justice Department allowed Google to buy a Boston start-up with the unprepossessing name of ITA Software for $700 million, but with an interesting caveat. The case is worth a closer look.

Five years after its 2006 founding in Boston, ITA probably had more data on flight bookings than almost any other service provider. It provided reservation and booking "back office" services for airlines selling tickets through online platforms like Bing Travel, Kayak.com, CheapTickets, as well as their own websites. With this expertise and $100 million in venture capital, ITA had conquered an interesting market into which a Silicon Valley data giant elbowed its way with Google Flights. Google's acquisition of ITA at first sounds like the familiar story of a start-up entering the Kill Zone: Goliath buys David so David can't become dangerous. But the US Justice Department approved the purchase with one significant condition: Google had to allow others, including (and especially) direct competitors, to access the acquired data.

The general public paid scant attention to the details of the Justice Department's condition. Experts in the field, however, saw it as an important and potentially influential signal for an idea that has since gained traction around the globe. We have already mentioned the push for wider access to data by the EU Commission. But the debate is also brewing at the national level, at least in Europe. For instance, in Germany, one of the ruling coalition parties has backed legislation to enforce comprehensive access to data within its manifesto. In the Netherlands, think tanks and policy makers alike are pushing for broader data access rules, partly in the hope this may afford the country a competitive advantage in attracting innovative entrepreneurs. Similar policy positions are also being discussed in the United Kingdom and Switzerland.

The US Justice department ruling when Google acquired ITA has helped inspire these discussions. The case is particularly instructive if one looks at how it ended. The access requirement was time limited. When it expired after five years, Google shut off free access the very next day. Overnight, multiple third-party vendors no longer enjoyed access to data for their services. There is an important lesson here: A data sharing requirement cannot have a sunset clause. Innovative entrepreneurs must be able to rely on continued data access. How? What legal hurdles have to be overcome? How can such access be technically implemented so that data remains secure, and sharing is both practical and inexpensive? The answers are sometimes less hard to find than skeptics fear.

Which Data? How Much? From Whom?

The first question is, what data should be included in an access mandate? Clearly only data that isn't subject to a legal obligation of confidentiality. In many jurisdictions, such obligations would apply not only to personal data, but also to trade secrets. Access to data is about nonpersonal data, or data that has been depersonalized. It's obvious therefore that access requirements don't compel nations to give up on data protection.

Hence, Google, for example, would have to grant third parties access to the one million most popular search query terms (and their most frequent misspellings), but not to Google account information or IP addresses of users. Amazon may have to share data about which products are sold, but not to whom. Neither aggregate search queries nor product names are confidential data, they are nonpersonal data points that can be shared. Critically, an access requirement would not grant third parties access to all data, only to a small subset. To promote competition even further, one could imagine linking the relevant percentage to the size of the data holder. Large collectors like Google or Amazon would then have to grant more access relatively speaking than small ones. Such an arrangement would deliberately favor start-ups, as well as smaller and medium-sized competitors.

In fact, small and medium-sized companies, including start-ups, benefit in a variety of ways from an access requirement. First, they gain the resources to transform their ideas into marketable innovations. Second, the advantage they draw is bigger than for large companies because major players already have a huge amount of data available. Every additional data point large companies gain from others is relatively less valuable to them than the same data point is to small and medium-sized companies. It's a bit like General Electric or Siemens as opposed to a small start-up being given $1 million in venture capital. GE or Siemens hardly notice, but for the start-up, it's the opportunity of a lifetime. Third, large companies have to share more data than small ones—an extra boost for the small guys. And finally, data from a variety of sources is particularly important for companies deprived of it because it tends to compensate for systematic errors in data collection, making insights gained from data more robust.

Requesting access to data would be available to all, independent of size. On the other hand, only companies of a certain size would need to open their data stores—for example, those with revenue above $25 million or with more than ten thousand customers. Such a setup would avoid burdening small businesses with additional responsibilities. But those small companies that request access to others' data would in turn be required to reciprocate. This would prevent selfish market players from being all take and no give.

These thresholds could be pegged to easily comprehensible parameters. Figures such as revenue or number of customers are easier to determine and verify than more complex metrics such as market share, for example (since it's sometimes unclear which market applies). This clarifies data access rules and prevents companies from gaming the system by playing down their market power or exaggerating the size of the market they're active in. In short, larger companies are required to grant access, and the larger they are, the more access they have to provide. The smallest companies can participate if they want, but if they choose to do so they would be required to give and not just take. This permits small firms to "grow into" access obligations when they are ready.

Critics of required access suggest such regulations could not be enforced worldwide. Would that leave the GAFAs (Google, Amazon, Facebook, Apple), the Big Four of Silicon Valley, free to continue business as usual? Of course not. Take GDPR as a case in point: Big Tech could not evade it by arguing their headquarters lie outside Europe's borders. Anyone who wants to do business in Europe must also submit to European standards. The same holds true at the national level. In early 2021, Australia passed a law that limited what local news content social media platforms like Facebook could disseminate without compensating news companies. Initially, Facebook threatened to cripple its platform for Australian members. But when the legislators did not budge and enacted the law, Facebook quickly signed licensing deals with Australian media. And these are just two of many possible examples: China forced Apple and Microsoft to conform to its rules, Turkey disciplined YouTube, as did France to eBay and Yahoo. Big

Tech might deeply dislike having to comply with laws they view as onerous to their business, but if the market at stake is large enough, they will comply. Access requirements would be no different. It's not just jurisprudence, but regulatory reality that demonstrates that nations can act here, as long as digital monopolists want the business.

Incidentally, the case of GDPR also underlines the upsides for Big Tech. At least to an extent, GDPR established trust for online interactions (and transactions) to thrive. In the context of data use, an access mandate would bring with it participation in an ecosystem of innovation. Big Tech would also benefit from access to others' data, albeit to a comparatively lesser degree. But the privilege of participation translates into significant leverage for regulators when enforcing requirements. If companies violate their obligations to grant access, they face expulsion from the data ecosystem.

Siloed Thinking in the Data Silo

There are a number of ways access mandates could be implemented institutionally and structurally. There are two key issues. First, will data be centrally stored and requested from that central unit, or will it be exchanged directly and locally between the data holder and the requester? And second, is access tied to certain requirements, such as the consent of others, or is it fundamentally open to everyone?

Policy makers steeped in traditional dirigiste and centralized policy may be tempted to establish an entire infrastructure for data sharing, which they could then administer—a huge data silo where third parties could retrieve data. But not every such megalomaniac fantasy needs to become reality, including this one. Simply put, central data silos are impractical. A lot of data isn't fixed. New data is added; existing data is changed or deleted. Each of these alterations would have to be transmitted to a central data silo, regardless of whether anyone has requested access to that data or not. That would entail continuously updating enormous amounts of data for no immediate reason or gain—a costly and gigantic effort for which there is no need. It would be even more problematic if companies could no longer decide for themselves where to store their data but were forced to choose a central data silo.

Moreover, there is no shortcut: Even though third parties get access to only a sample of the data, that sample has to be randomly selected for each request, which means that all data of all companies has to be present in the central data storage for it to work. A central data silo also increases data security risks because it's an attractive target for hackers from all over the world. Once they had successfully hacked their way in, a nation's collective data treasure trove could be plundered.

The opposing model uses decentralized transmission between data requesters and data holders. This is both more practical and less expensive, as companies continue to manage their data themselves. They transfer data to meet a specific request, not in advance. This lowers data transmission overheads. The decentralized model also has advantages regarding cyber security. Because it is decentralized, a breach will have limited consequences, encouraging overall resilience.

In discussing the advantages and disadvantages of centralized and decentralized storage, some point out that data in a central silo would, of course, be encrypted and thus safe from hackers. In addition, "data fiduciaries" could oversee data management and control access. But both these measures would help only to a limited extent because, irrespective of encryption, a gigantic and valuable data silo represents risk. Hacking incidents in recent years of supposedly securely encrypted data storage should serve as a warning when it comes to claims of complete security. A new profession of "data fiduciaries" won't solve this risk problem, neither objectively from a technical-organizational perspective nor subjectively in people's minds. Who will be the fiduciaries? Why should we trust them? Does this mean that major telecommunications companies would get control over hordes of data—or an army of lawyers with keen legal noses, but virtually no technical expertise?

Some commentators have suggested that only the government (whether national, state, or local) can be a suitable data fiduciary. But in many nations, the government's digital competency doesn't provide much reason to hope that it could successfully take on the major technical role of running such a data silo, even apart from the costs that this setup would entail. A decentralized model in which data is transmitted directly from holder to requester is more resilient, leaner, and more efficient—and thus quite simply better. And it requires no new intermediaries, in whatever guise.

Give and Take

The second big decision when it comes to an access mandate is whether access will be open or closed—that is, whether basically everyone will be allowed to participate or only a select group. In technical jargon, the question is whether the aim is a closed data pool or an open system of data access. GitHub, the world's largest platform for open-source code, now owned by Microsoft, is pretty well accessible to everyone. Anyone can take part by downloading source code or uploading it to GitHub. It's a largely open system. The California Data Collaborative, on the other hand, maintains a pool of cleaned, standardized, and enriched water-use data accessible only to its members, mostly local water utility agencies in California.

Policy makers are grappling with this choice. For instance, in February 2020, the EU Commission announced it would develop or promote the establishment of so-called data spaces for a range of application areas and economic sectors. If they are limited-access data pools, they would contradict the open, decentralized spirit that Saxenian saw as partly responsible for Silicon Valley's record in innovation over several decades. By the same token, EU Commission president von der Leyen spoke clearly of the need for open structures. So here, as in many similar contexts, the jury is still out as to the actual structural and institutional arrangements for broader access to data that will be chosen, although logic seems to suggest a preference for an *open and decentralized strategy*.

But in concrete terms, how would a request for data access work under such conditions? It would require a regulatory framework that was a shade more comprehensive, but fortunately that can be implemented relatively easily and at low cost. The main issue is how to discover who holds what data, how to select data, and what data a requester receives. The first part could be achieved by making each company's existing databases more transparent so that competitors know who in general terms has data about what. If you don't know where to look, a right to access won't do you much good. The goal is not to make available a detailed and comprehensive description of all data, but rather a rough division into two or three dozen standard categories such as Internet search, product recommendations, training data for speech or image recognition, geographical data, and the like. Every company that's obliged by its size to grant access to others would be required to maintain an entry in a kind of telephone directory for data access, update it once a quarter, and indicate for which of the two or three dozen standard "buckets" it has data on hand. The entry would also include the Internet address where a request can be lodged, and what percentage of data will be made available (which could be, as we have suggested earlier, determined by the size of the company and a handful of "brackets"). The directory would be accessible to everyone online. With its help, companies could decide who to request data from.

The request itself would be transmitted online via standardized interfaces, directly linking requester and data holder. There is no need for a data trustee, a central data silo, or government involvement. This simple, immediate, and direct system was already proven when Google had to grant access to others in the wake of its ITA acquisition (see earlier discussion). It also means that no one will have to transfer data to a central point in advance or continuously update it in anticipation of a request. Legal rules would determine how quickly a request has to be answered and limit the number of requests per year (although access to frequently updated data would be possible more often than access to data that rarely changes). Here too the aim is to create simple, transparent, and cost-effective solutions that are fit for purpose without overly burdening the companies providing access.

Each request would specify which category of data is being requested. However, the company requesting access wouldn't be able to select data more narrowly than that. Neither would the data holder; it would have to provide a new, randomly selected data set for each request. This ensures a fair subset of the data would be made accessible. It's also important that the access mandate does not specify how access to data is to be implemented technically, so that the development of technical tools and innovative solutions to fulfilling access mandates can proceed unhindered.

The only substantive requirement on data is that it can't be "naked," but must include sufficient meta-information about the nature of the data being transmitted. For example, if data stems from a conventional database, metadata will entail information about the individual data fields and their origin. For this too, there are established standards employed by many data holders, similar to code that represents a standardized article number. Or take temperature data, as an example. It would necessitate metadata about what was measured and according to which scale. And that's it. There's no need for complex procedural steps during access, and also no need for new institutions. Government's role would be narrow and simple: it would operate the aforementioned online directory, but deliberately avoid acting as an intermediary. It wouldn't get involved in the practical day-to-day workings, but ensure compliance with—and if necessary, enforcement of—the access mandate.

The basic principles of access to data are straightforward and transparent. The focus is on efficient implementation for all participants, with significant impact upon competition and innovation. To enable small and medium-sized companies to make use of such data access opportunities, some additional measures may be necessary. For instance, smaller companies could be loaned expertise, perhaps even receive subsidies, as they build up in-house technical and organizational expertise. This begins with the task of collecting and storing data in appropriately standardized form so that in exchange, robust insights can be gained from the analysis of diverse data sources. Digital giants invest hundreds of millions of dollars annually in developing their own data standards, which all their business partners then have to adhere to. This too represents an attempt by Big Tech to secure and extend their informational power. Therefore, it's incumbent upon national and international industry and standard setting institutions in the relevant sectors to establish and maintain open standards accessible to all.

Although we see wider access to data applying only to nonpersonal data, companies are often uncertain whether data sets contain personally identifiable information and thus whether data protection rules apply or not. And they are struggling at times to reliably select nonpersonal data from a larger set, or to strip data from personal identifiers, so essentially to de-personalize it. But here, too, progress is being made. Data protection authorities have worked on guidelines for effective depersonalization, and technologies have been developed to make depersonalization easier and more robust. But more can and should be done.

It may also require adjustment of liability rules to ensure that data holders are not held responsible for how data they have been mandated to make accessible is being used by third parties. Because data users reap the economic value from the use of data, it is only fair to hold them, rather than original data holders, responsible.

Data portability—an individual's right to oblige data processors to transfer one's personal data to third parties and included in recent data protection laws like GDPR—is of little use in this context. If it is to have any impact at all, many individuals would have to act in lockstep. In practice, however, this very rarely happens. Political scientists call this the problem of collective action.

Similarly, it would make just as little sense to insist on some form of data "property," and have, for example, Facebook pay people a few dollars each year for their data. That would have no effect on the overall power structures. Minor fixes will accomplish nothing against the in-built advantages of the data giants. On the contrary, when it comes to data use, the small fry can learn a lot from the bigger fish.

Benefitting from Reuse

From the data that Apple cars collect for its map app, the company also gains information that enables it to determine more precisely the geographical location of iPhones. Google can use data from its intelligent NEST thermostat to predict energy consumption. From sensor data supplied by its server farms, Amazon gains highly detailed knowledge of how long particular hard drives last. Netflix crunches user data to shape the plots of its next series. Internally, the data-wealthy superstar companies are masters of data reuse. But all of these new insights are generated in-house. With a broader access to data, data reuse would happen on a much larger scale and across corporate boundaries.

Today, many traditional companies are unable to reuse their own data at all. And those who do are often only looking for a confirmation of the familiar. They use their data to answer questions they've already posed instead of asking new questions inspired by patterns in the data. Their direct competitors frequently are no better; they see the world from the same vantage point. This is precisely why broader access to data leads to a strong stimulus for innovation: new users with entirely different ideas gain access to the data.

There are countless examples of this. Earthquakes can be detected using data from underseas fiber optic cables normally used to transmit data, and meteorologists can improve weather forecasts using sensor data that passenger aircraft collect to feed their autopilots with. Logistics experts can use maintenance sensors in cargo trucks and trailers to optimize loading processes and fleet utilization—a substantial benefit, because almost one in three truck trips is an environmentally harmful empty one. From the constantly changing price information on online platforms, reusers of data are able to calculate daily inflation rates. And car navigation data has been used to model commuter streams far better than before. Reuse of data is particularly successful in the field of biogenetics, where for instance it yielded breakthroughs in the development of new antibiotics.

With open data access, much more would be possible. The reuse of auto telemetry data could identify dangerous sections in the road network. Machine data from intelligent networked factories could provide important information for the development of new materials. We'd finally be able to learn where exactly 3-D printing would be advantageous and where not. In health care, individual reuse of data would enable us to predict more accurately when certain drugs would be effective for certain patients and at what dosage, and where in our surroundings there may be a particularly high risk of infection. And in human resource management, the reuse of data would make it possible to predict not only who would be good for a certain job, but also happy in it. There are already pilot projects under way in these areas as in many others, but in almost every case, they were initiated and carried out by data users outside the sector.

The bottom line is that seemingly useless data becomes valuable when it's used to answer new and different questions. But these questions are mainly being asked by third parties and outsiders. While data monopolists may have mastered the game of data reuse, an open access mandate raises reuse to a completely new level. Of course, this comes at the expense of some of the monopolists' relative market power—but at the same time offers benefits to many, many more.

Open Data and Open Minds

Politicians around the globe have responded to the rise of Big Tech with proposals to make it difficult or even impossible for Silicon Valley behemoths to buy up local digital start-ups. But such a mercantilist, and deeply defensive stance will hardly help. It's important to keep Schumpeter's nightmare in mind. Stopping acquisitions in themselves won't get rid of the structural advantages digital superstar companies enjoy. It's far better to link consent for the acquisition of an innovative digital start-up to stringent data access requirements. Access mandates such as the one imposed on Google in the ITA case, but without the crippling term limits, could kill two birds with one stone: Long-term access to data would strengthen other market participants. And conversely, data giants would lose some of their informational power.

But perhaps the most important precondition for data sharing is not a matter of laws and regulation. The critical factor is the mindset toward data—how important it is to make good use of it. Only if companies understand that data is the key to innovation will they appreciate mandated access to data for what it is—a once in a lifetime opportunity that boosts competition and turbocharges their innovativeness. But this requires decision makers to realize that data has no value in itself—its value accrues only through using it. That seems obvious. But corporate practice tells us data is hardly used. Executives in many companies have yet to grasp the profound importance of data use in the digital age.

We often equate the Industrial Revolution with the invention of the steam engine. But that's wrong. "The" Industrial Revolution never took place as a single event. Instead, there were multiple waves of industrialization. And the most important one wasn't even the steam engine. Instead, it was the effective electrification of factories. When electric motors began appearing in America's factories in the late nineteenth century, they replaced steam engines. But nothing else changed at first. Just as with steam engines previously, a large central electric motor powered all machines in a factory through myriad drive belts. That brought little gain in efficiency and a lot of risk. If the single motor broke down, the factory ground to a halt.

It was only when people realized that the production *process* could be changed thanks to electric motors that manufacturing was transformed. The motors got smaller and were used directly where power was needed on the factory floor. If one of the many engines broke down, the factory as a whole no longer came to a standstill. Manufacturing became more efficient and resilient. But this required less of a technological leap than a rethink, a change of mind, a different attitude. And making that mental shift is challenging—not only for companies, but society as a whole.

Today, the situation is similar: we need to change our mindset. We can only gain insights from data if we use data widely and repeatedly. Our societies are on the cusp of a new and necessary stage of learning. Experts often speak of "data literacy." This doesn't mean that everyone has to train to become a data scientist, let alone learn to understand the complicated prediction models that let "quants" extract insights from mountains of data. Of course, we need more of these experts with outstanding statistical skills, but for the bulk of us, "data literacy" entails above all a mindset that makes the best of wider access to data.

That means the understanding and constant willingness to utilize data for value creation and societal innovation. Data literacy and a wider access mandate will mutually reinforce each other. This could unleash a dynamic like that of Silicon Valley in the second half of the last century. The timing could not be more propitious. Over the past decade, vertically integrated structures and closed thinking have led Silicon Valley to increasingly resemble Route 128 shortly before its descent into irrelevance. The information flows that enable the digital giants' power and dominance rest on exclusive access to data that is neither economically rational nor socially legitimate, since it thwarts innovation.

Nations around the world have the opportunity to make data more useful for everyone through a logical complement to their national information privacy data laws. Such a mandate for data sharing, as previously set out, would not only break up the power asymmetries between Big Tech monopolists and users freely handing over their data. It could also help an even grander vision turn into reality: open data.

7. *Open Data Reloaded*

It's late December 2019, and cases of severe respiratory infections are increasing in the Chinese city of Wuhan. On Christmas Day, doctors send a sample taken from patients to a laboratory that specializes in the genetic sequencing of bacteria and viruses. The result comes three days later: the sample contains a new member of the coronavirus group. Chinese and Australian researchers work around the clock to determine the virus's genetic code. Only two weeks later, on January 10, 2020, they publish the entire gene sequence and make it available on the Internet to the global research community. This enables every public and private organization on the planet to develop medications or vaccines to fight the virus. Just seventy-two hours later, the crucial protein, developed by members of the US National Institutes of Health, is ready for the vaccine platform of American biotech start-up Moderna. Then extensive and lengthy testing begins as part of the approval processes that all vaccine candidates are required to undergo. Whether a vaccine is truly safe and effective cannot (yet) be tested by computer simulation.

By the end of 2020, the first vaccines were approved for use, and population-wide vaccination commenced, first focusing on particularly vulnerable groups in society. Compared with any vaccine development in history, even most recent ones, the pace of development for COVID-19 vaccines was breathtaking, compressing time to develop and test from about a decade to less than a year.

On the political level, especially for relations between the United States and China, the pandemic appears to have deepened rifts. In the world of science, however, it has increased openness and constructive collaboration among researchers. Confronted with the pandemic, the international research community opted to let others access the results of their work. They would openly share data they have collected. We hope this will become one of the key lessons of the pandemic, and that it will shape the way such data is treated from now on: as a public good for accelerating innovation for the benefit of all. That government can facilitate openness and access is exemplified by another major example from the field of life sciences: the Human Genome Project (HGP).

The HGP began in 1990 under the coordination of the US National Institutes of Health. The goal was the complete sequencing of a human DNA, all 3.2 billion base pairs. It was set up as a cooperative project among international research institutions. The collected data, it was agreed, would be made available promptly. Ten years and more than a billion dollars later, the HGP announced it had achieved its goal (though not all the genetic information it had collected had been put in the correct order yet and made public).

But the HGP researchers weren't the only ones on a genetic mission. A private US company, Celera, had begun deciphering human DNA on a large scale in 1997 using a new technical approach. Celera also promised to make most of the data accessible, but expressly reserved the right to patent individual decoded gene sequences and thus secure exclusive use for itself and potential licensees for a limited period of time. At that point, US president Bill Clinton intervened to persuade Congress to enact an explicit ban on the patenting of human gene sequences. As a result, human gene data remained accessible without any bar. No company, the thinking went, should be able to secure a monopoly or even a temporary exclusion right to the data that describes our bodies biologically and defines how we develop and function.

As a consequence of the act of Congress that made the human genome a public information asset, American biotech companies lost around $50 billion in market value within a few weeks. But today, hardly anyone remembers. The biotech industry's current business models are no longer based on monetizing the possession of human DNA information, but on using the data. And that has led to a huge dividend, not just for biotech companies but for society as well. To give but three examples: Using HGP data, researchers discovered drugs already developed to treat other diseases could also be highly effective against several types of cancer. HGP data led to the development of new and simpler diagnostic methods for diseases such as hepatitis. And freely available DNA data enabled the development of a drug to treat the respiratory disease cystic fibrosis. With this drug, made available in 2020, over 90 percent of cystic fibrosis patients, until then forced to endure tremendous suffering and premature death, should enjoy a normal life.

The HGP's approach to information, based on free access to research data, is now a widespread guiding principle in life sciences in general—for example in the Genomic Data Commons of the US National Institutes of Health, a data pool of genetic information on cancer, or the European Nucleotide Archive, a comprehensive global collection of molecular data.

Open Data Renaissance

The development of vaccines against COVID-19 and the HGP are examples of information sharing and collaboration. And this reinforces a larger trend based on the conviction that the insights gained from data should benefit everyone. The idea of open data is enjoying a renaissance. Its origins date back to at least the 1950s when geophysicists agreed on data standards to make information exchange easier. NASA advanced the idea in the 1970s in the context of satellite development and remote sensing data. Beginning in the 1990s with the World Wide Web, global initiatives urged governments to make much of their data available online. In many countries, this push led to new laws and initiatives, although mostly limited in openness and modest in scope.

Around that time, American Beth Noveck came to Europe to study historical examples of how the erosion of trust in government could push societies to authoritarian rule and worse. For Noveck it was an intellectual as well as personal journey. Her roots were European, and she had been brought up with European culture and values. But through her family ties, she also was all too familiar with the unspeakable suffering caused by the Nazis and their totalitarian allies. As she was sifting through primary sources, Noveck came to appreciate the central importance of open information for the well-being of societies, especially democratic ones—and of how a lack of openness and transparency may destroy social trust and put a society onto a slippery slope toward dictatorship. Her historical studies taught her the value and importance of good governance, and why openness furthers and facilitates it. As she returned to the United States and became an academic, the connection between societal trust and openness turned her into an open data activist and a role model for many, seeing a new function for open data that greatly transcends notions of public sector efficiency. It also set her on course to shape open data's trajectory in the United States and abroad—and become an influence for other open data activists who we'll meet later in this chapter.

For many open data advocates, data is fundamentally a common resource that should be accessible to all. At first glance, this may sound radical, but when you think about it, it's far less so. The free flow of information and the availability of facts are the lifeblood of democracy. Without them, people wouldn't be able to ground their opinions in evidence and fulfill their democratic role. Without them, we would have a concentration of information in the hands of a few, threatening not only innovation and competition but also democracy and freedom. As Benjamin Franklin realized at the dawn of the American republic, open access to information is a necessary tool to empower the people. It's only through open information that citizens can play an active role in democracy, create and maintain the lively public sphere that sustains the democratic venture, and (our immediate concern) ensure that they receive a fair share of the digital dividend—not as a handout, but as the necessary consequence of the important role in society they play.

Many of our legal systems deliberately guarantee not only the fundamental right to freedom of expression, but also the right to receive information without undue restrictions—a right to access information. And where access is restricted, for example due to copyright or data protection rights, the restriction is narrowly defined as an exception. This is backed up by international treaties, such as the UN's International Covenant on Civil and Political Rights. Ratified by over 170 nations from around the world, the

Covenant states in Article 19 that the right to freedom of expression explicitly includes the "freedom to seek, receive and impart information and ideas of all kinds, regardless of frontiers, either orally, in writing or in print, in the form of art, or through any other media of his choice." Simply stated, the principle of open access to information is not extraordinary or radical but a fundamental component of the legal and societal fabric of many nations.

Therefore, the crux isn't the principle of open access to information, but its concrete implementation. In the sciences, open access to data is already pretty much a reality. A long list of countries has widened access to government data. Nongovernmental organizations (NGOs) around the world are successfully collecting and making data available to all. And more and more companies are assuming their responsibility by giving society access to evidence and data so that individuals and communities can make better decisions. In the following, we'll take a closer look at these dynamics. Sneak preview: The sciences have made considerable progress along the path of open data, but most governments still have a long way to go when it comes to access to data. Some NGOs are surprisingly far advanced along this road. And the desire of some large companies to share more information gives some cause for hope, even if the returns so far are comparatively paltry.

Open Access: When Scientists Refused to Play Along

Open data in research is a story of success—and of liberation. At the close of the twentieth century, scientific publishers had deftly established themselves as crucial intermediaries. They amassed big monies for their role as exploiters of information, even though many doubted the value they truly added. While researchers published their results without a fee, would-be readers had to shell out occasionally shocking amounts for access. Annual subscription rates in the thousands of dollars for a single science journal were not uncommon. Research information became heavily commercialized. What's more, in practical terms, it was also only accessible to the few organizations around the globe that could afford these usurious rates. Smaller research institutions in poorer countries had virtually no chance of competing. In the 1990s scientific publishers began offering their content on digital platforms, much like the digital giants today. And just like in Silicon Valley, an unprecedented process of market concentration led to a small handful controlling most of the world's high-quality research publications. This created information silos and imposed artificial limits on engaging in research and innovation among many societies. But in sync with the transformation of publishers into highly profitable information platforms, a powerful counter-movement was forming: open access.

Many researchers were no longer content to be humble servants of a scientific publishing oligopoly that restricted access to their own published work. Researchers *want* to be noticed, and this is best achieved with broad open access. For their part, government funding outlets became increasingly aware they were paying publishers vast sums for public universities and research organizations to access the results of research they had funded. The situation called for a revolution, and the resulting movement for open access in research has become a much-vaunted success story. Thanks to low-cost digital distribution, thousands of open access journals sprang up online. Almost half of all research publications worldwide today are freely accessible in one form or another, around double the percentage from fifteen years ago. Three of every four university-published science journals are open access. In South America it's as high as 80 percent, while in Western Europe it's only 25 percent. The European publishing oligopolies are still engaged in a defensive battle. But policy makers are turning against them. New funding guidelines insist researchers publish their results open access. And, for example, in the United Kingdom universities receive state funding only in proportion to published research results that are available through open access, even outside of specific project funding. The core of the movement's demands is often summarized with the acronym FAIR. It stands for "findable, accessible, interoperable, and reusable" information. Both the EU Commission and the G20 have thrown their support behind FAIR. The trend is obvious—and in the natural sciences, it has only intensified in the wake of the COVID-19 pandemic.

In fact, it's no longer just a matter of free access to research *results* but also to the underlying *data*. It makes sense, as real value stems from using it. So, there is even less reason to keep research data under wraps than there is to keep the lid on research results. The G8 announced its support for "open science data," even before the pandemic, while initiatives are on the rise worldwide to make public research funding conditional on the publication of collected data.

Of course, the sciences are not completely open access. Envy, zero-sum thinking, and ambitions of power aren't completely unknown phenomena among scientists. But the scientific world is profoundly rethinking how it operates, especially with respect to access to data. More and more researchers are coming to accept that the raw material of their publications has to be accessible, not least to ensure that their work is reproducible and verifiable. They understand that other researchers will be able to use this data to answer different questions or analyze similar questions in a different way and thus obtain new scientific insights, which don't necessarily compete with their own work. It is also a consequence of the responsibility to serve and benefit science and embrace it as a collaborative project. In addition to this focus on scientific ideals and values, the technical capabilities for sharing information have also improved dramatically. In the past, this was often a major technological challenge. Today almost all research data is recorded digitally and can be easily shared over the Internet. As standards are becoming more firmly established, data can also be more easily classified and analyzed by third parties. The inspiration for open access to research data comes from one of the most data-rich research institutions worldwide.

Close to Geneva, on the border between France and Switzerland, lies CERN (the European Organization for Nuclear Research), the world's largest center for particle physics. Since 1952, researchers from all over the world have been working on answering the question, what fundamentally holds the world together? It is a joint project of twenty-three countries. In addition to more than three thousand permanent staff, CERN hosts more than fourteen thousand visiting scientists from over eighty nations. CERN is not only the world's leading center for nuclear research, but also a symbol and institution of open and unhindered access to information that has embodied this spirit for decades. It's no coincidence that this is where British scientist Tim Berners-Lee came up with the idea of the World Wide Web in the early 1990s, transforming the Internet into a global network for information exchange available to all and sundry.

What's particularly fascinating is the organizational structure Berners-Lee chose for the Web: not central data silos, but a delibberately decentralized structure of distributed information that could be easily linked together from anywhere in the world. The result was a worldwide virtual fabric of information that anyone could access. The Web allows us to combine and share information on a global scale by easing much of the technical work of requesting and assembling individual pieces of info. The web's qualities—scalable, simple, dispersed, and therefore resilient—made it the greatest technological disruption of recent decades. Its structure wasn't imposed by technology, but a conscious choice of technological design. Working at CERN, Berners-Lee set up a decentralized network because he wanted the world to have diverse and open flows of information. Because that's what he saw and experienced among his colleagues at CERN every day. It's not surprising that CERN has been one of the largest providers of research data in the natural sciences. It has made accessible more than two petabytes—that's a number with fifteen zeros—of such data.

Open access is no longer limited to individual areas of science such as medicine. Large and valuable data repositories for physics, chemistry, biology, geography, and meteorology are just as accessible, as is an increasing amount of data from the social sciences. For instance, if you want to know how certain values have changed in various societies in recent decades, you can go to the World Values Survey and access valuable data sets collected over as many as forty years from dozens of countries: online, open, simple, and comprehensive.

Not only research institutions, but also international organizations are becoming aware that their task is to make data about the world openly available to all. The World Bank, the OECD (the Organization for Economic Cooperation and Development), and Eurostat, the provider of statistical information at the European Union, are just a few of these beacons of data access. And if you want to save yourself the trouble of finding the right data source on the Internet, you can use Google's Public Data Explorer, which also provides tools for quick analyses of data sets.

The COVID-19 pandemic boosted international information cooperation. In March 2020, as the number of infections in Europe and North America were skyrocketing, open and freely accessible data sources sprang up like geyser plumes. Within days, researchers at Johns Hopkins University in the United States had made available a tracker of infections and deaths worldwide, which accessed open data sources. And if you were at all skeptical about official death figures, you could turn for example to the CDC (Centers for Disease Control) in the United States and EuroMOMO (European Mortality Monitoring Project) in Europe for stats on excess mortality. All these examples demonstrate how quickly salient information could be provided thanks to freely available and openly accessible data sources. Statements can now be checked for their accuracy—and not just by a few Ivory Tower experts with exclusive access to relevant data.

As far as we've already progressed in the sciences, there's still much more to be done. In some areas, such as medicine, earth sciences, or archaeology, better data standards are required so that data from different sources can be more easily combined. National and international infrastructures for research data need to be expanded more swiftly and comprehensively. In doing so, it's important to keep the focus on decentralized and scalable structures and resist the temptation to build large, centralized data silos that would likely be expensive, risky, and cumbersome. But fundamentally, science is clearly blazing the trail in providing more open access to data and adopting the corresponding mindset. Here, public and private sectors can learn a lot from scientists.

GovData: How the Public Sector Can Enable Innovation through Data Access

When one hears of open government data, one may think of pollution data, or details on public transport demand, of subsidies and their recipients, or perhaps of data from a country's central bank. But the most dramatic success story of open government data actually is from the US military. In the 1970s and 1980s, the United States put a constellation of satellites in place, with which soldiers could pinpoint their and their enemy's location with high precision worldwide. The system known as GPS proved to be crucially important for the US victory in the first Iraq war.

GPS's real global impact came in 1983 when a Soviet fighter jet accidentally shot down a Korean passenger airplane that had strayed from its course. In the aftermath, US president Ronald Reagan announced that GPS, once operational, would be made available to civilian users around the world—a pledge that was expanded by President Clinton in 2000. It jump-started innovation in aviation and shipping, in logistics and fleet management, but also in consumer-level car navigation and most of our smartphones. The free stream of location data gave birth to a large ecosystem, and the free precise time signal that came with it enabled a wide spectrum of additional uses, from the financial industry (time-stamping transactions) to broadband mobile and Internet connectivity. The immediate market size for GPS-related products and services is estimated to exceed $100 billion annually, with a strong growth trajectory. GPS brought safety and resilience to ships, airplanes, and cars. It facilitated better decision-making for organizations and communities. And it enabled a sea of new products and services and led to countless new market opportunities.

But at its core, GPS is little more than sensitive military data that one government, the United States, decided to offer to the world, not necessarily as a gift, but as a down payment to enhance safety and stimulate innovation. And GPS is not alone. A 2013 study by James Manyika and colleagues at the McKinsey Global Institute estimated the total additional economic value generated worldwide in just seven sectors through the intelligent use of open data, particularly from governments, could be around $3 trillion.

This estimate may be fairly recent, but the underlying idea is not. Access to government information has been a topic of discussion in Western societies for decades. The principle of confidentiality prevalent in national bureaucracies long barred access to government data. But the principle is neither immutable nor the only possible approach. In Scandinavia, open public records—the exact opposite of bureaucratic confidentiality—is a centuries-old tradition. It hasn't slowed economic and societal development. In fact, northern Europe is flourishing, both informationally and economically.

The importance of a transparent public sector in sustaining democracy, it is worth remembering, became shockingly apparent when in the 1970s US President Nixon instructed employees to break into a campaign office of his opponents and cover it up. The Watergate scandal led to Nixon's resignation. It was only exposed because a confidential source inside Nixon's administration leaked information to journalists from the *Washington Post*. This underscores the need for democracies to be able to control government bureaucracies through transparency and access to information, and to be able to counter their informational power when necessary.

In the United States, as a direct lesson from the Watergate cover-up scandal, Congress passed both a data protection law constraining federal authorities and a thorough revision and drastic strengthening of federal freedom of information rules. But access to information had to be requested and was granted on a case-by-case basis only. Often, government agencies refused access and had to be sued in court to eventually relent and reverse course. This limited the practical availability of information. Worse, it established freedom of information as an adversarial process, in which the role of government agencies was to resist rather than assist the information seeker in her quest.

Freedom of information needed to move out of its narrow adversarial confines if it was to have an impact. A first step in that direction was taken two decades later, under President Clinton. The Electronic Freedom of Information Act (E-FOIA) required American federal authorities to proactively open their data troves and make data digitally accessible online. By moving beyond an adversarial process, freedom of information was reestablished as an integral duty for government agencies. It was the dawn of modern open government data.

But, as Beth Noveck recognized, it wasn't nearly enough. While technical hurdles persisted and freedom of information rules needed further clarification, the biggest impediment to open government data was structural. As long as lawyers and bureaucrats were in charge of implementing open data, little would change. For open data to really take off, responsibility for it would have to shift to government's chief information officers, who could appreciate the value of information to society and who had the expertise and resources to turn open data into a reality.

Having suggested an intriguing open data approach to combat a crippling backlog in the assessment of US patent applications, Noveck, now a rising star in the global open data community and an established professor in New York, came to the attention of the early presidential campaign of a young senator from Illinois, Barack Obama. She played a key role in drafting Obama's open data chapter in his election platform. After the election she joined the administration at the president's science and technology office in charge of open data, drafting an executive memorandum instructing government agencies to make public sector data accessible by default, which with some luck, as she explains, became one of the very first acts signed by the new President. This greatly aided open data to land in chief information officers' portfolios.

Today, digital access to official data is opening up in many countries. But upon closer look, and unlike in the early Obama administration, one realizes that embracing open government data has not yet fully arrived in the daily administrative practices of many government agencies around the world. Data is routinely kept inaccessible with the argument that it could contain personal or other confidential information even when there is no personal data at stake. And government agencies still employ too infrequently simple depersonalization measures that could make many additional data sets available.

Sometimes the principle of open government data fails because of the complexity of bureaucratic layers in a modern state. Federal and state agencies may have sufficient expertise to make data available, but municipalities face real hurdles. Even larger cities can be overwhelmed by the challenge of setting up and maintaining the infrastructure for open government data while small villages very often fail to take even initial, basic steps toward transparency. More importantly, a mindset steeped in the need for confidentiality still runs deep, letting government officials often opt to share as little as possible. However regrettable, this is an understandable reaction. For centuries, the power of bureaucracies has been built on information advantages. Throughout their professional lives, public officials were told about the need to maintain confidentiality. It would be presumptuous to believe that a political desire for open data could swiftly turn around a deeply ingrained bureaucratic culture. What's needed is a pretty fundamental shift in mindset. That is why open government data requires more than just investment in technical and organizational infrastructures for data sharing; it needs education and training in data literacy among public sector employees. Even that, however, won't suffice. Empirical studies, including in countries where many public sector data sets have been made accessible, have shown that the use of data falls far short of expectations. There are two main reasons for this.

First, interviews with developers of open data applications have exposed that these developers often lack any sense of being connected to a community. Hence, it's not enough for government to work on data supply. It also has to look at the demand side and create the conditions for a data ecosystem of app developers to take root. In a way it's similar to Apple's App Store. It too is an ecosystem in which Apple deliberately took on the task of doing more than just making the developers' apps easy to find and install. Apple also boosts developers' success via information and networking initiatives; it uses marketing and PR to draw users' attention to new and innovative apps. Open government data also needs such an ecosystem, and this requires information and communication—for example, to let developers know when new data sets are becoming available so they can better and more reliably plan their work. Also helpful are opportunities for networking, as well as a channel to publicize the best apps.

Noveck, working hard for open data in the Obama White House, was well aware of this need. She knew that she needed to show government CIOs how open data can actually be released in practice, and she needed to show data analysts and NGOs and the general public what can be done with open data to whet their appetite. So she and her colleagues designed and built an online platform to access US open data as well as highlight new applications and novel data analysis based on the data. It opened the flood gates. Within months, hundreds of thousands of data sets became available online and were easily accessible using the data.gov platform and its search functions.

Second, public sector data sets by themselves are rarely interesting. They usually enable new insights only in combination with other data sources. As long as these other data sources remain inaccessible—for example, because they are held by companies as proprietary information—open government data cannot develop its full potential. That's why open government data without mandatory access to data held by private sector entities is like a rider without a horse—ambitious but confused and destined for failure. Conversely, because open data adds value through the use—that is, the analysis—of data, it also needs to be possible for the private sector to make use of open government data for their own purposes, including commercial ones, without having to hand back part of their income—for example, as license fees. Open government data is primarily an opportunity to ensure transparency and create value for society. But if done well, it can also turn into a huge donation of data to the economy and stimulate innovation. More and more political decision makers are realizing this. So, it's particularly good news that the commitment to open government data is increasingly going beyond the narrow limits of administrative data.

For instance, beginning in December 2019, the European Union required public transport companies to make their timetable data digitally accessible to the public. The goal is innovation in the mobility space so that a wide variety of entities (and not just Google) will be able to provide travel information—not just locally, but across the entire continent. This is another case of data-driven innovation that both adds economic value and creates benefits for society, as it helps people act more sustainably by choosing public transport more frequently. Less encouraging, however, is that compliance with these legal requirements is patchy so far.

As of mid-2021, many public transport companies in Europe are not adhering to these data sharing requirements. In particular, the foot-draggers are unwilling to disclose real-time data. They argue that this doesn't constitute raw data but is instead processed data outside the scope of open data guidelines. Here we find yet another example of a misguided mindset—withholding data because you don't want to let third parties use it. It needlessly squanders an opportunity for innovation through competition, while the data holders—in this case, the companies—are hardly in a position to offer effective and comprehensive curb-to-curb mobility information online. The result is not just an economic loss. Society, too, loses as the lack of easy-to-use mobility information limits the appeal of public transport.

The principle of open access to data has the potential to be a powerful lever. It can improve access to research data, but also to data from both the public and the private sector. Boosting this leverage is undoubtedly sensible, but this requires not only improved regulations, but also, above all, consistent enforcement. Unlawfully withholding data is not a trivial slip, but an intentional slap in the face of innovation and a denial of digital dividends for us all. Those public agencies that restrict access to data are behaving even more unscrupulously than the monopolists which control commercial data; they are deliberately disregarding existing laws. It's a public responsibility—shared by political decision makers, the media, institutional watchdogs, and vigilant citizens—to call them out. The lawbreakers need to develop an awareness of the injustice they cause.

But this also suggests we need more than rules and regulations. Public sector bodies that are being asked to open up access to data need our help to enter the age of data—organizationally and logistically, but above all by evolving the right mindset. We can't demand open government data, for example, and then fail to give municipalities the technical tools and human resources to meet this requirement.

Donating Data

The sciences and government agencies are just two areas where access to data is being broadened. There are also NGOs that collect data and make it available to the general public.

Wikipedia is perhaps the best-known example. A little less well known is OpenStreetMap, which manages a digital world map. Over six million users have helped build this free and open alternative to Google Maps. It contains more than forty million buildings and data for more than 80 percent of all roads worldwide. What's most exciting is that OpenStreetMaps data can be used not only as a freely accessible world map, but also as raw data for analysis. This can translate into, for example, better aid coordination after natural disasters, such as when Hurricane Dorian devastated the Bahamas in autumn 2019. Or data that can be used to analyze the development of urban districts and neighborhoods.

But this just scratches the surface of what's on offer.

A particularly interesting initiative has been undertaken by the Mozilla Foundation in the United States, which developed the Firefox browser. In its Common Voice project, Mozilla is collecting data donations in the form of recorded human voices. The goal is an openly accessible training data set with a wide variety of voices and accents to broaden the reach of speech recognition systems. Common Voice has already collected well over one hundred thousand voice samples and converted them into digital data in dozens of languages. It focuses on diversity so future speech recognition systems will successfully recognize linguistic minorities.

The real surprise, however, is companies that donate data voluntarily. After science, government, and NGOs, open data has finally reached the C-suite. Some providers of satellite data, for example, are offering free access to data on the world's forests. It can be used to identify slash-and-burn clearing, but also areas where forests are in grave danger for other reasons. Data donations are not limited to environmental data. For instance, one of South America's largest real estate companies opened access to data on property values so that dangerous price spirals and real estate bubbles can be identified as soon as they emerge.

And even the otherwise ravenous data giants of Silicon Valley sometimes make some of their data accessible to society. In April 2020, in the middle of the first COVID-19 wave, Apple and Google made available mobility data, showing human traffic in various contexts (retail, public transport, sports, bars, and restaurants). Lockdowns became immediately visible—and also where and when social distancing measures were no longer being observed. This provided valuable insights for policy makers. At the same time but independent of the pandemic, Microsoft announced a comprehensive open data push that includes both data analysis tools as well as access to valuable data, for example on the use of broadband networks. And Facebook's "Data for Good" initiative provided analytics tools to enable public decision makers as well as society at large to make evidence-based decisions during the pandemic.

All of this is helpful. The open data initiatives of some of the superstar companies are evidence of a new sensibility among their executives. They are becoming increasingly aware, often because of pressure from their own employees, that their informational power entails an elevated level of responsibility to society, and that their actions up to now have often fallen short. Some of these executives may perhaps be thinking tactically: better to donate a little data and generate positive headlines than to attract even more attention from policy makers. And of course, these voluntary data donations from Microsoft, Google, Facebook and other Big Tech firms are at best a drop in the ocean. But how about thinking far more radically, and imagining a world of open data? A world in which scientific, public, and private sector data is largely accessible to all?

Open Data World

In a world of comprehensive access to data, start-ups as well as medium-sized companies would have significantly better prospects of turning their ideas into innovations. Davids who come armed with more data could more often take on the Goliaths and win. Digital laggards would no longer be able to hide behind the argument that, *of course*, they were fully committed to digital transformation, but unfortunately, due to a lack of data, they just couldn't move forward any faster with it. Barriers to entry into digital markets would be lowered and fresh wind would blow in the sails of the imaginative, the enlightened, the agile, and the hard-working. Reinvigorated competition would reduce outsized monopoly profits, lower prices, and create real choices for consumers.

Open access to data would also mean access from anywhere. It would translate into new opportunities for smaller cities and rural regions—and the inventors and entrepreneurs there. In a world of open data, we would gain new insights and innovations not only where data led to large, fast, and easy profits, as in online trading, online marketing, or online gambling. Democratizing data will make data accessible to those who hope to advance the causes of sustainability, environmental protection, and the transition to renewable energy. Data for everyone will improve political decision-making and help elevate political discourse. Although it seems hardly conceivable, many decisions today, involving everything from public transport to education to health care, are made on the basis of very little and barely meaningful data. Which new subway line is necessary, which school most urgently needs to be modernized, and where more nursing staff should be deployed—with open data, all these questions could be deliberated on the basis of factual evidence and decided more transparently. Open access to data would therefore not only promote democratic processes and ensure that we use scarce resources appropriately, it would also allow us to keep a closer watch on those in power and encourage them to use more data in reaching their decisions.

In our digital world, there's a vast gap between technical possibilities and social reality. We have platforms and tools to make information easily available to others. But as data is hidden in jealously guarded silos, digital monopolies act in concert with predigital economic stragglers and a public sector that loathes change. It should come as no surprise that so far "social" information sharing tools have been used primarily for the exchange of trivialities or to promote political polarization. In contrast, real open access to data could empower citizens to exchange ideas with each other on the essential issues of the day and make more *informed* decisions.

This vision of open access to data does not mean the end of data protection. There would be rich information dividends for businesses and society just by making available nonpersonal nonconfidential data. But we must not forget: each act of empowerment also creates a possibility for abuse. We have to learn to act accordingly. In our present state of digital overload, we often agree to comply with the general terms and conditions of digital apps without thinking, and privacy and consumer legislation suggests as long as we consent, that's okay. But it's not. The burden of informational responsibility rests on the wrong shoulders. In a world of open data, responsibility should be borne by those who derive most of the economic benefit from it, and cases of abuse should lead to severe penalties. Or bluntly put: data protection should no longer regulate access, because so far that has only served to facilitate the concentration of informational power. And that in turn has routinely led to corruption, oppression, poverty, and war.

A world of open data is no land of milk and honey; the data dividends won't just fall from heaven like manna. It's an opportunity, but someone will still have to do the work. Opening access to data is at best a necessary, but not sufficient, prerequisite for rekindling the fire of innovation, as well as restarting a more robust and thoughtful public discourse and facilitating a democratic renaissance. We have come to a decisive crossroads. If we continue down the path of exclusive access to information, we will strengthen data silos, boost digital platforms, and make our democracy structurally vulnerable to information manipulation, corrupt power politics, and irrational, fact-free populism. Access to data is the primary lever to empower people by reversing the insidious dynamics of information concentration. And let's be clear: this means empowering people not just in the West, but all over the world.

8. *The End of Data Colonialism*

A woman in Chicago types a destination into her Uber app and schedules a ride. Uber's algorithms set the price and find a suitable driver nearby. Confirming the ride takes a few seconds longer than usual. What the customer and the driver don't know: Uber's real-time ID check has sounded an alarm. The driver just started his shift and was required to upload a selfie to the system. The night before, however, he shaved because it was his girlfriend's birthday. Without a beard, the facial recognition software wasn't sufficiently certain whether the man behind the wheel was in fact the Uber driver he said he was. At the same time, a young woman in Hyderabad, India, is working at her laptop in her kitchen. She works for CrowdFlower, a platform for so-called clickworkers.

Two pictures of the driver appear on the laptop's screen, the official one from his Uber account with a beard, and the selfie that he's just taken. An hourglass starts to count down the seconds. The woman in India has only a few moments to decide if it's the same man, now clean-shaven. She authorizes the ride with a click and earns a few cents for it in return. Anthropologist Mary Gray and computer scientist Siddharth Suri describe this scene in their 2019 book *Ghost Work: How to Stop Silicon Valley from Building a New*

Global Underclass. For Gray and Suri, "ghost work" is one of the many emerging, startling, and paradoxical phenomena that digitization has brought about. Billions of people around the world use digital services on their computers and smartphones and believe that intelligent machines run these services. But in fact, hundreds of millions of people around the world are completing tasks that have been broken down into their smallest possible packets in return for minimal payment, and thereby keep the giant digital machine's wheels turning. Even Frederick Winslow Taylor would probably be astonished how well his concept of the division of labor scales on the assembly lines of data capitalism, how well a single worker's output can be measured and monitored, and what economic benefits clients can gain from the disaggregation and partial automation of work.

The platforms that broker this work through micro-contracts, invisible to users, have trendy start-up names like Cognizant, Streetspotr, Figure Eight, Mechanical Turk, or—dropping any pretense of euphemism—Clickworker. These platforms exist because still today, when we think we engage with AI, often there is a real human doing the work: Comparing a clean-shaven face with a bearded one, recognizing and tagging traffic signs in video recordings made by Waymo's cars, writing mini-product descriptions for online shops, researching telephone numbers and entering them into databases, cleaning and preparing data sets, counting football players' ball contacts, screening content for violence, hatred, or pornography on social media, thinking up trivial answers to the trivial questions people ask Alexa—humans are still better at these tasks than machines. This doesn't suggest that AI will not soon be able to take over. The more training data is collected and used, the likelier machines will replace humans in these roles too. But today and at least in the short term, even superstar firms remain dependent on human prep work. And with the clickwork platforms, they've created a system in which the individual human day laborers are likely to feel much more dependent on their clients than the other way around—especially considering the increasing role AI will play in substituting human clickworkers.

In their book, Gray and Suri describe the helplessness of freelancers scattered across the globe who are forced to compete for micro-jobs. The globally distributed "cyberiat" is of course in no position to unionize like the workers in the wake of the Industrial Revolution. Pay for such work, as researchers-cum-activists like Oxford's Mark Graham note, is well below the minimum wage in the Western world, thanks to a merciless price war.

A research group led by Kotaro Hara at Singapore Management University and Abigail Adams at Oxford University calculated an average hourly wage of $2 for participants on Amazon's Mechanical Turk clickwork platform when unpaid time spent searching and preparing for each task was factored in. Only 4 percent of clickworkers earn more than $7 an hour. In many countries, minimum wage laws, even if they exist, are not enforced when it comes to work done on micro-contract platforms. And, of course, platform providers almost never contribute to social security and health care for their clickworkers. They insist that from a legal point of view, clickworkers are self-employed agents who automatically enter into a contract for each small task. The measly remuneration is typically credited to their account through a global payment service such as PayPal.

Given these working conditions, almost nobody in a prosperous economy will find it worthwhile financially to toil as a clickworker; at most, it's a way to pass the time in return for a little bit of extra income. That's why the overwhelming majority of digital day laborers are sitting at their computers in India, Pakistan, Bangladesh, the Philippines, and in Latin America, particularly in the countries of Central America as well as Brazil, Peru, and increasingly also in destitute Venezuela, as long as electricity is working and the Internet connection is good. In Africa, too, many people are clicking away for $1 to $3 an hour. For many people in less affluent regions, individually this may be an economic opportunity, better than the options on their local job market. From a societal perspective, however, this development radically contrasts with digitization's grandiose promises of emancipation and the resulting "flat world" that Thomas Friedman described some twenty years ago.

Today, we are experiencing a rebirth of colonialism. This renaissance is increasingly referred to as "data colonialism," a term popularized by the Mexican-American scholar of communication studies Ulises Ali Mejias and the British sociologist Nick Couldry. This time, the people of the Global South aren't being exploited by gunboats, land grabs led by colonial armies, subjugation and brutal violence, or the predatory exploitation of natural resources. In data colonialism, the former colonies instead provide data through clickwork or by using apps. And the colonial powers aren't the United Kingdom, France, Italy, Spain, Portugal, Belgium, the Netherlands, Germany, or Japan, but the United States and China. For the first time, the former European and Japanese colonialists are in the same boat as those whom they exploited historically. Of course, most Europeans and Japanese haven't been compelled to resort to clickwork by dint of their precarious economic situation. On the other hand, having to accept the general terms and conditions of the data colonial powers' apps bears some resemblance to the treaties that colonial rulers had powerless representatives of the peoples of the South sign before they stole their land and their resources. Clicking okay doesn't mean accepting a deal agreed by equal partners, let alone understanding its arbitrary terms and conditions.

Tech Cold War

In 2018, Ian Bremmer, head of the Washington-based think tank Eurasia Group, suggested a label for the geopolitical standoff between the United States and China over digital technology: "Tech Cold War." Bremmer's argument is as follows:

Since the end of the East-West conflict, the world has lived under a Pax Americana, not only politically but also technologically. Apart from a few niche areas, US hardware and software has dominated the world market. Internet start-ups in Silicon Valley networked the world, used the resulting data more intelligently than anyone else, and rose to global superstar status, more valuable than all other previous companies. The American political sphere proudly observed the worldwide triumph of American technology. Policy ambitions were largely limited to preventing regulation from getting in the way of the global proliferation of US technology. US foreign trade policy was in sync with the interests of the tech companies and the drumbeat of globalization.

By the same token, everyone could read in each of China's new five-year plans how the country was strategically orchestrating its path toward becoming a Tech Nation. How well these plans were succeeding via "capitalism with Chinese characteristics" could be seen in the achievements of Chinese tech brands such as Lenovo, Xiaomi, and Huawei, first in the huge domestic market, and later on the global scene. For data-driven applications from shopping to payments, for mobility services, social media, digital assistants with speech recognition capabilities, or health applications, Chinese products are now often superior to their American counterparts.

With the rise of China to digital superpower status, the technological Pax Americana has slowly faded away. As in the Cold War between NATO and the Warsaw Pact, in Bremmer's view, the world is now increasingly being divided into two camps. And countries will have to choose to align with one superpower or the other in order to use that power's digital platforms and technologies. According to this scenario, the networks of the United States and Chinese technospheres will become decoupled from each other. Places where these two digital systems collide will see skirmishes between surrogate forces, as in the Cold War between the Americans and the Soviets. And as in the years from 1945 to 1989, military and political loyalty will go hand in hand with economic ties, which in the "Tech Cold War" will be reinforced by hardware, software, and digital infrastructure. A quick look at tech headlines seems to confirm Bremmer's binary view. Anyone operating a 5G network with Huawei technology is pressured by the US camp. Those who opt for Chinese technology but can't afford it can get inexpensive loans from Chinese state banks.

Since at least 2020, the situation saw further escalation. In its last months in office, the Trump administration expanded the political battle zone from chips, smartphones, and telecommunications infrastructure into the realm of data by pushing for the divestiture from the Chinese owners of social video app TikTok, with more than one hundred million mostly young users in the United States, and by threatening a ban on social app-for-everything WeChat, highly popular among many Chinese-born Americans.

The political style of the noisy though ineffective saber rattling of the Trump White House may remain a historical exception. But the Biden administration, while de-escalating the toxic rhetoric, actually toughened some key policy stances. It sees China as the real geopolitical competitor and high tech as the proxy battlefield. With plans for an enormous investment in R&D and an invitation for aspiring young researchers from around the world to come to the United States, the Biden White House hopes to improve its position in the battle for technological leadership and global market share in the high-tech sector, especially in data-rich services.

Driven by policy considerations, both sides will try to at least partially decouple their supply chains. The Chinese will increasingly develop their own chips, and American hardware manufacturers such as Apple, HP, and Dell will relocate part of their production to other countries, if only to reduce their exposure to geopolitical risks. Moreover, recent regulation from China's central government restricts the use of foreign-made hard- and software in government agencies. In the battle for supremacy in AI, China has a structural advantage thanks to its huge population and their enthusiastic use of digital services. Data in particular will remain a hot policy issue. The Great Firewall of China blocks America's digital champions, and if Europe had any, they too would probably be denied access to the Chinese market. Until recently, Chinese superstar companies have been allowed to conquer markets in the West despite serious doubts about the security of data collected by them and the potential of sustained data access by Chinese government agencies. TikTok and WeChat are but the initial skirmishes over data supremacy in the Tech Cold War. But more and more countries are considering excluding Chinese suppliers, such as Huawei, from key roles in the emerging mobile data supply chain.

Let's hope that this Tech Cold War, if we accept that it is one, doesn't heat up, as Harvard professor Graham Allison suggests it could in an ominous scenario laid out in his latest book, *Destined for War.* In it, Allison points out that whenever in world history a dominant world power shows signs of decadence and decline and has been challenged by an emerging great power, it has almost always led to war. In reference to the Greek general who led the Athenians against Sparta, Allison calls this danger the "Thucydides trap." An example in more recent history is the rise of the German Empire under Bismarck and Wilhelm II, which increasingly challenged British global hegemony, ultimately leading to World War I.

Skeptics of the historically loaded image of a Tech Cold War, such as the Indian-American author and political advisor Parag Khanna, point out that the division of the world into two hostile technological blocs contradicts the essence of digital technology. Tech spreads around the world without limits and at warp speed. To a skeptic, it seems neither technologically nor politically feasible for two superpowers that are increasingly unpopular in the rest of the world to erect a digital Iron Curtain. That may be true, but Europe, Japan, the emerging Asian economies, and the Global South still have every reason to abandon the role of passive spectators in the geopolitical struggle between the two digital superpowers. Naturally, the metaphors of a Tech Cold War and data colonialism can be challenged not only on linguistic grounds, and, of course, they greatly—and perhaps dangerously—simplify digital reality. But especially when taken together, the two dystopian visions offer indications about the relative digital powerlessness of the rest of the world, but also how infusing a geopolitical dimension into the high tech sector greatly complicates matters for United States and China as well.

The danger is that the emergent tech cold war, much like its historical antecedent, develops into a dynamic that is difficult to stop and step back from. Seen through a geopolitical lens of hegemony, both the United States and China will perceive the need to win to retain and enhance their respective positions of power.

China, the emerging digital power, may see its role as challenging the incumbent technology hegemon to better project its own rise to power. In part, it may also be a reaction of US threats to keep domestic the key nodes of the high-tech value chain, such as technology design and development as well as cutting edge production and data hosting. To extend its hegemonic reach, China is already offering its putative digital client states an appealing—at least at first glance—package of economic and infrastructure aid, access to inexpensive products, and, for rulers whose grip on power is experiencing some turbulence, political stability through economic growth stimulated by China. China has also been increasingly injecting its interests in international standard setting institutions and processes, often taking leadership positions in them to influence the trajectory of technology's development arc.

The United States, on the other hand, may perceive itself as the incumbent threatened by the newcomer, prompting a defensive stance and a focus on retaining every inch of its global high-tech position. The United States is responding with protectionist measures and renewed demands for loyalty from its allies in digital and data policies, particularly but not exclusively when security issues are involved. Every nation that chooses a Chinese supplier, every standard setting process that doesn't go the US way, may be seen through the lens of a zero-sum-game, in which one side's win, however insignificant it may seem at first, is only a stepping-stone toward greater geopolitical might on the tech war battlefield.

These and similar processes may lead to a self-sustaining dynamic in which the two geopolitical contenders jockey to divide and colonize the entire digital world between them. In fact, United States and Chinese Big Tech companies already have injected themselves into most corners of this new world, the digital *terra nullius*. They have created informational power structures in which smartphone users export the digital economy's most valuable raw material directly to the servers of the data colonialists in exchange for the small change of access to free services. Their systems aren't necessarily optimized to benefit the users, but to entice them to spend more time using them and thus maximize the amount of data they generate. These digital colonial grandees decide not only how the data streams will flow from European pandemic tracing apps, but also what billions of people in Europe, Latin America, Africa, and Asia see in their personal news feeds, and who can and can't eavesdrop on their emails and data streams. In what is perhaps the most darkly ironic element of this information power game, the ultra-wealthy data colonialists with countless billions of dollars in their corporate bank accounts pay the clickworkers of the Global South paltry wages to generate precisely the training data the machines will need for their next step: automating the "ghost work" of the clickworkers.

In the Tech Cold War era, it's time for a new movement of data decolonization.

Digital Decolonization

Whenever using a term as historically tainted as colonialism, it would be wrong and unacceptable not to recall the endless suffering that colonizers caused through the oppression, enslavement, and genocide of Indigenous tribes and peoples in subjugated colonies. And here of course our analogy to the digital age reaches its conceptual limits. Yet, the checkered history of political decolonization, especially after World War II, also may help us better identify where potential pitfalls lie in any attempt to remove the power of data colonialists.

Much like their historical counterparts, Big Tech in the United States and China has achieved three things. First, these Goliaths have created technical structures in which the economic and (to an extent) political power relationships have become structurally entrenched so that the appropriation of resources leads to even greater dominance. Second, the continuous appropriation of data as a resource and the exclusive use of this data are made to appear the completely natural order of things and thus legitimate. Digital users rarely rise up against the status quo, as if, viewed from within, there is no alternative to the market economy's globalized logic of exploitation. In fact, this appropriation of data is so much taken for granted that no one even notices the process of expropriation. And third, the data-rich superstars have created a narrative for their powerless data suppliers that sounds as crude as the historical colonizers' claim—that the colonized peoples should in truth be grateful that the colonialists finally brought them "civilization" and that they are part of a great movement toward progress. As we saw in the first chapter, the heroes of Silicon Valley are master storytellers. Their Chinese colleagues have shrewdly mimicked this ability as well. This time the world is supposed to find deliverance from its problems through a digital panacea that can only be developed in Silicon Valley and China, as the rest of the world is unfortunately unable to advance digital innovation fast enough.

Political decolonization and the dynamism of postcolonialism that continues to this day demonstrate how enduringly powerful colonial power structures can be. Oppression backed by military force was replaced by postcolonial economic exploitation of labor and raw materials, this time backed by predatory contracts enforceable in international courts. In the former colonies, the Indigenous elites who came to power often ruled as accomplices of the former colonial masters and against the interests of the (formally) independent countries. For generations, the self-image of the people in the colonies was at least partially shaped by the negative external image that the supposedly civilized white colonial rulers drew of the colonized peoples, endlessly drumming a sense of inferiority into them through White schools, White literature, and White media.

All of this seems like a grossly exaggerated version of the way that many nations in Europe, Asia, Africa, and Latin America are being chained down by data colonialism, and yet, as they become increasingly aware of their predicament, they still try in vain to maneuver their way toward digital sovereignty. Discussions are often dominated by a mixture of an acute sense of inferiority and forced optimism. Policy makers, traditional companies, and other actors in regional digital economies hope to be able to catch up somehow in data capitalism's global race for market power. Their faith is in being able to do the same things just as well, or at least almost as well, as the ruling monopolistic platforms. Numerous nations talk a good game about "digital sovereignty" and national "champions," but prefer to organize yet another study group or resort to old-fashioned industrial policy subsidizing their local digital laggards instead of expanding access to data and instilling the appropriate data mindset, the prerequisites for developing AI systems and digital applications in the making.

Major emerging markets sometimes react allergically to the most direct attempts to seize digital power, as India did to Facebook's proposal to provide free Internet access for the whole country (as long as Facebook could be the gatekeeper and tap into all the data). But by limiting their policy strategy to infrequent and incoherent demonstrations of power, these populous countries with their countless users and wealth of data are, in effect, reinforcing information inequality just like the seven hundred million Europeans, while an even higher proportion of their top tech talent emigrates to the United States.

The Coalition of Data Colonies

In chapters 5 and 6, we described why a comprehensive data access mandate could kickstart digital and innovation liberation, and what its legal and practical implementation could look like. This option is not only available to advanced postindustrial regions like Europe, but to every country and every jurisdiction with a sufficient number of users and market power that superstars wouldn't want to leave to others.

In April 2019, the United Nations Conference on Trade and Development's week-long forum on digital development in Geneva discussed a proposal for digital decolonization with the help of a data access mandate, in precisely the same location where the League of Nations' 1919 charter took the first steps toward historical decolonization in Africa and Asia. The confident feedback from delegates of major emerging economies including India, Pakistan, and Nigeria was, we can enact a data access mandate more rapidly and comprehensively than Europe with its complicated decision-making processes and complex data protection regime. This is not only factually accurate, but if it were undertaken, it would also send an unmistakable political signal of the Global South's self-empowerment in the age of data-driven innovation.

But the best solution would be for many regions to converge and form a data access coalition to take on the data colonialists. Perhaps surprisingly, initial steps toward such a vision have already been taken. In June 2019, Japan's then prime minister Shinzo Abe launched an initiative at the G20 summit in Osaka to promote a more open exchange of data internationally. Its express goal, referred to as the "Osaka Track," is to promote innovation and competition through international data cooperation. The Osaka Track was endorsed by twenty-four countries, including many European ones as well as the United States and China, the data superpowers. What was striking, however, was that data-rich emerging economies such as India, Indonesia, and South Africa refused to join in. Their sense is that Europe only wants to use plans for international data access to build a third digital power center and that then, following the logic of the Tech Cold War, the Global South would no longer have to choose between two digital hegemonic powers but three. The rocky start of Abe's "Osaka Track" is powerful evidence that perhaps the days of digital hegemony by United States and Chinese Big Tech companies are numbered.

A strategy for creating wealth through accessible data across borders could entail a lucrative offer to many nations: become part of a data access region based on carefully considered and consistently enforced rules; within this region, US and Chinese companies will have to grant access to their data if they want to do business. This isn't a conventional postcolonial concept. Quite the contrary: It's meant as a blueprint to disrupt rampant hegemony-preserving postcolonialism through joint action. If this endeavor of forging a new alliance of such nations is successful, it could fundamentally change the power structures that the superstar companies have established. Informational power could become decentralized and distributed throughout the world. That sounds like a technological solution, but it's actually structural and thus far bigger than that: it would be a revolution for global society; digitization could finally fulfill its grand promises of emancipation.

The advocates of digital systems wanted to strengthen democratic discourse, create economic opportunities and prosperity for everyone, and show emerging economies and the Global South a path of progress in a technologically "flat world." At times we seemed to be on the right track. But an increasing amount of evidence is pointing in a different direction. Political discourse is becoming radicalized almost everywhere, with superstar companies tweaking algorithms to fuel the flames, since conflict increases user engagement, and time spent engaging with social media can be monetized. Economic inequality is growing almost everywhere in the world, both within societies and between nations. And, as we described in detail in chapters 2 and 3, despite all claims to the contrary, we aren't living in innovative times. Innovation is slowing down because of market concentration. What's often overlooked is that too little innovation fosters social division. Only radical technological and social innovation will create the qualitative gains in prosperity that will enable those who are economically excluded today—in both hemispheres—to participate in the future, and at the same time help humanity to get our planet's ecological problems under control. Elon Musk is welcome to proclaim that he wants to die on Mars. In a certain sense, that's even admirable, since among the well-known figures of Silicon Valley, Musk is the only Schumpeterian entrepreneur: a creative destroyer who hopes to blaze a trail for something new to come into the world (or into outer space) through vision, purpose, and single-minded implementation. But most of us don't want to die on Mars. We want to live a good life on Earth.

Access Rules

As the complex history of decolonization illustrates, successfully breaking up established power structures is an absolutely necessary but also hugely daunting task. But the twentieth century offers an illuminating, if often overlooked lesson of how this can be done. In the 1950s, US telecommunications company AT&T and its various subsidiaries and affiliates had become a corporate behemoth. It employed a million people and raked annual revenues in excess of $30 billion in today's terms. But these figures belie its true power. Ma Bell (as it was affectionately called) manufactured the telephones, operated the switchboards, and ran the US telecommunications network. It provided the wires to connect nuclear missile sites with the Pentagon and the Pentagon with the White House. It owned and operated the undersea cables the connected the US with the rest of the world. What to the individual customer may have looked like a phone company was nothing less that the monopoly provider of the lifeblood of communication for the world's leading economic superpower.

But AT&T was even more important than that. Hidden away in the New Jersey suburbs of Murray Hill, AT&T operated Bell Labs, its vast corporate research center. Engineers at Bell Labs were tasked with research and development to improve and evolve AT&T's phone network. But they were given a long leash from headquarters, and in the process invented an avalanche of key technologies far beyond voice communication. It's most important and most consequential invention, however, was the transistor in 1947. It is quite literally the foundation for the entire information revolution. Leaders at Bell Labs quickly understood the enormity of the invention and made sure they patented it right away—as well as a slew of processes to manufacture transistors from silicon. In some ways, Bell Labs and AT&T were the equivalent of all of Silicon Valley compressed into one corporation.

AT&T's enormous economic and information power also brought it into the crosshairs of the US government's antitrust division. In 1949, the federal government sued AT&T for abusing its monopoly position. After a protracted legal process and sustained lobbying (including stressing AT&T's importance for national security at the height of the Cold War), the government settled with Ma Bell in 1956. AT&T and its subsidiaries were regulated by local public utility commissions, but AT&T remained largely untouched—with one seemingly small, but hugely momentous exception: almost all patents awarded to Bell Labs by 1956—about eight thousand—would be available to any American company for free; and all future patents granted to Bell Labs would be available to companies for a small fee.

With one stroke of a pen, the entire intellectual basis not only for telecommunications networks but also for the information revolution fell into the public domain, providing the US economy an information and knowledge dividend of unprecedented scale. And it was a gift that kept giving for many decades—as transistors turned into the building blocks of computer chips, and as even without the exclusivity of full patent protection, the staff at Bell Labs kept discovering, developing, and designing things, such as the laser, digital image sensors, mobile phone technology, as well as the C programming language and the Unix operating system (whose descendants today power billions of computers, from supercomputers to smartphones).

The case of Bell Labs is important and revealing in at least three ways. First, it demonstrates the positive impact such an open access decision can have to the wider economy—and to the dynamic of innovation. Recently, researchers have shown that the opening up of Bell Labs's patents increased follow-on innovation, particularly outside telecommunications. And this activity was driven by small and medium-sized companies. As we mentioned earlier, the co-inventor of the transistor and long-time Bell Labs researcher William Shockley offers a telling tale. When he left Bell Labs in 1956 for his own chip start-up in Menlo Park, California, and quite literally jump-started Silicon Valley, he could rely on thousands of Bell Labs patents for extra thrust.

Second, the case offers evidence that companies which have to open access to others aren't becoming less creative or inventive. Bell Labs's researchers kept producing important discoveries and inventions well after the consent decree that settled the antitrust case in 1956. Of the nine Nobel prizes awarded for work at Bell Labs, seven were for discoveries after the consent decree.

Third, the United States has successfully used open access mandates to spur innovation with significant economic and societal benefits. If the Republican Eisenhower administration at the height of the Cold War can force a shockingly powerful monopolist to open access, nothing prevents policy makers in the United States and elsewhere to do the twenty-first century equivalent—except a lack of understanding of how access rules can turn into a game changer leading to long-term benefits for society.

Access to data won't correct these overnight. But if we disrupt the monopoly power of superstar companies by opening access to data, innovative entrepreneurs throughout the world may have the same opportunities to improve our collective lot by using decentralized digital engines of empowerment. On a level playing field, in a Schumpeterian spirit of disruptive creation and with a pro-competition mindset, they will bring about digital innovations that benefit everyone.

Is that an idealistic utopia, or a feasible vision? Benjamin Franklin's answer was clear: access to information is the most important basis for democratic discourse, economic development, and political justice—the point of departure and the driving force of the American Revolution. It resulted in the independence of the United States from British colonial power. Data for everyone is the beginning of the end for data colonialism. Access rules. When everyone has access to the informational riches of the data age, the nature of digital power houses will change. Information technology will find its way back to its original purpose: empowering us to use information so we can live better as individuals and as societies.

Acknowledgments

This book grew out of earlier works we wrote together, including "Reinventing Capitalism in the Age of Big Data" (Basic Books 2018) and in particular "Machtmaschinen" (Murmann 2020).

We are most thankful to our agent Lisa Adams of Garamond Agency, who has been a sage advisor. We thank Tim Sullivan, who heads the University of California Press and was the editor for Viktor's "Delete" more than a decade ago, for welcoming our treatment of an important, but complex subject. We thank Michelle Lipinski for her most thoughtful guidance, outstanding conceptual eye, and unfailing support. Her professionalism and enthusiasm, and that of her colleagues at the Press, has been hugely motivating and deeply inspiring. We are lucky to have found such an exceptional home for *Access Rules.*

We gratefully acknowledge the help of Jonathan Green for translating an earlier version, as well as David Gow and Lynda Crawford for their superb editing skills. We also thank two peer reviewers, who have read and commented on the manuscript—and whose feedback we have tried to incorporate.

We are thankful to the many people who have spoken with us about various aspects of this book, on and off the record. *Pars pro toto,* we'd like to thank explicitly professors AnnaLee Saxenian and Beth Noveck, who rightfully play important roles in the narrative we offer.

Thomas Ramge would also like to thank the Weizenbaum Institute for the Networked Society and the Center for Advanced Internet Studies for enabling him to do part of the work on this book during his research fellowships there.

With good reason, authors tend to end the acknowledgments to thank their loved ones. We want to stick to this tradition—not so much because it is right (it is!), but because we are acutely aware of the fact that as we added word after word in front of a screen, life went on. And our families stoically accepted that we were working on yet another book. We can't promise it's the last—but as we look to the future that our children will inherit, we strongly believe in the significance of this book's message.

Further Readings and References

The argument in *Access Rules* transcends a single academic field and weaves together strands from a variety of disciplines. For readers, who are interested in the works we found useful, we offer the following noncomprehensive list. The references in the first section cover more general works that are touched on in multiple chapters. They are a great starting point for anybody wanting to gain more of an overview perspective. The references for each chapter offer extensions to the more concrete points we put forward in each chapter; these are ideal for deep dives and for readers with an interest in a specific dimension of our argument.

Much has been written about digital platforms, how they threaten democracy and undermine our society. Many such works have focused on issues of information privacy, fake news, and the manipulation of public discourse. Others have looked at platforms from the vantage point of critiquing capitalism. *Access Rules* aims to offer a different perspective, linked to innovation but also to openness, diversity, and values of the enlightenment. The suggested readings that follow reflect this perspective, not as a disregard to those other points of view, but as a complement to them.

General Reference

Brin, David. *The Transparent Society: Will Technology Force Us to Choose Between Privacy and Freedom*. Cambridge, MA: Perseus Books, 1998.

Castells, Manuel. *The Rise of the Networked Society*. New York: John Wiley & Sons, 2011; originally published 1996.

Christensen, Clayton M. *The Innovator's Dilemma*. Cambridge, MA: Harvard Business Review Press, 1997.

Coase, Ronald H. "The Problem of Social Cost." *Journal of Law and Economics* 3, no. 1 (1960): 1–44.

Cohen, Julie. *Between Truth and Power: The Legal Constructions of Informational Capitalism*. Oxford, UK: Oxford University Press, 2019.

Erixon, Frederik, and Björn Weigel. *The Innovation Illusion—How So Little Is Created by So Many Working So Hard*. New Haven, CT: Yale University Press 2016.

Fagerberg, Jan, David C. Mowery, and Richard R. Nelson (eds). *The Oxford Handbook of Innovation*. Oxford, UK: Oxford University Press, 2005.

Friedman, Thomas. *The World Is Flat: A Brief History of the Twenty-First Century*. New York: Farrar, Straus and Giroux, 2005.

Gordon, Robert. *The Rise and Fall of American Growth: The U.S. Standard of Living since the Civil War*. Princeton, NJ: Princeton University Press, 2016.

Khan, Lina M. "Amazon's Antitrust Paradox." *Yale Law Journal* 126 (2017): 710–802.

Khan, Lina M., and David E. Pozen. "A Skeptical View of Information Fiduciaries." *Harvard Law Review* 133 (2019): 497–541.

Mayer-Schönberger, Viktor, and Kenneth Cukier. *Big Data: A Revolution That Will Transform How We Live, Work, and Think*. London: John Murray Publishing, 2013.

Mayer-Schönberger, Viktor, and Thomas Ramge. "A Big Choice for Big Tech." *Foreign Affairs*, September/October 2018.

Mayer-Schönberger, Viktor, and Thomas Ramge. *Reinventing Capitalism in the Age of Big Data*. London: John Murray, 2018.

McAfee, Andrew, and Erik Brynjolfsson. *Machine, Platform, Cloud: Harnessing Our Digital Future*. New York: W. W. Norton & Company, 2017.

McNamee, Roger. *Zucked: Waking Up to the Facebook Catastrophe*. New York: Penguin Books, 2019.

Noveck, Beth. *Wiki Government: How Technology Can Make Government Better, Democracy Stronger, and Citizens More Powerful*. Washington, DC: Brookings, 2010.

Pariser, Eli. *Filter Bubble: What the Internet Is Hiding from You*. New York: Penguin, 2012.

Ramge, Thomas. *Postdigital: Using AI to Fight Coronavirus, Foster Wealth, and Fuel Democracy*. Hamburg: Murmann Publishers, 2020.

Ramge, Thomas. *Who's Afraid of AI: Fear and Promise in the Age of Thinking Machines*. New York: Experiment Publishing, 2019.

Saxenian, AnnaLee. *Regional Advantage: Culture and Competition in Silicon Valley and Route 128*. Cambridge, MA: Harvard University Press, 1996.

Shapiro, Carl, and Hal Varian. *Information Rules: A Strategic Guide to the Network Economy*. Cambridge, MA: Harvard Business Review Press, 1998.

Stucke, Maurice. *Big Data and Competition Policy*. Oxford, UK: Oxford University Press, 2016.

Syed, Matthew. *Rebel Ideas: The Power of Diverse Thinking*. London: John Murray, 2020.

Teachout, Zephyr. *Break 'Em Up: Recovering Our Freedom from Big Ag, Big Tech, and Big Money*. New York: St. Martin's Press, 2020.

Vaidhyanathan, Siva. *The Googlization of Everything (And Why We Should Worry*, Berkeley: University of California Press, 2012.

Véliz, Carissa. *Privacy Is Power: Why and How You Should Take Back Control of Your Data*. London: Bantam Press, 2020.

Weizenbaum, Joseph. *Computer Power and Human Reason: From Judgement to Calculation*. London: W. H. Freeman & Company, 1978.

Wu, Tim. *The Master Switch: The Rise and Fall of Information Empires*. London: Atlantic Books, 2012.

Zuboff, Shoshana. *The Age of Surveillance Capitalism: The Fight for a Human Future at the New Frontier of Power*. London: Profile Books, 2019.

Chapter 1

Abboud, Leila, Joe Miller, and Javier Espinoza. "How Europe Splintered over Contact Tracing Apps." *Financial Times*, May 10, 2020.

Benkler, Yochai. *The Wealth of Networks: How Social Production Transforms Markets and Freedom*. New Haven, CT: Yale University Press, 2006.

Boerding, Andreas, Nicolai Culik, Christian Doepke, Thomas Hoeren, Tim Juelicher, Charlotte Roettgen, and Max V. Schoenfeld. "Data Ownership—A Property Rights Approach from a European Perspective." *Journal of Civil Law Studies* 11, no. 2 (2018): 323–370.

Bria, Francesca. "Digital Sovereignty for the People in the Post-Pandemic World." Medium, August 24, 2020. https://medium.com/@francescabria/digital-sovereignty-for-the-people-in-the-post-pandemic-world-109472dd736b.

The Economist. "The Coming Tech-Lash." November 18, 2013.

Frasca, Ralph. "The Emergence of the Early American Press." *Pennsylvania Legacies* 6, no. 1 (2006): 11–15.

Isaacson, Walter. *Benjamin Franklin: An American Life*. New York: Simon & Schuster, 2003.

Lewin, Amy. "Barcelona's Robin Hood of Data: Francesca Bria." *sifted*, November 18, 2018. https://sifted.eu/articles/barcelonas-robin-hood-of-data-francesca-bria/.

OECD. "Why Open Science Is Critical to Combatting COVID-19." May 12, 2020. https://www.oecd.org/coronavirus/policy-responses/why-open-science-is-critical-to-combatting-covid-19-cd6ab2f9/.

Reidenberg, Joel R., N. Cameron Russell, Alexander J. Callen, Sophia Qasir, and Thomas B. Norton. "Privacy Harms and the Effectiveness of the Notice and Choice Framework." *I/S: A Journal of Law and Policy for the Information Age* (2015): 485–524.

Solove, Daniel. "Introduction: Privacy Self-Management and the Consent Dilemma." *Harvard Law Review* 126 (2012–13): 1880–1903.

Tufekci, Zeynep. *Twitter and Tear Gas: The Power and Fragility of Networked Protest*. New Haven, CT: Yale University Press, 2017.

Weber, Max. *Economy and Society: A New Translation*. Cambridge, MA: Harvard University Press, 2019.

Chapter 2

Agrawal, Ajay K., Avi Goldfarb, and Joshua Gans. *Prediction Machines: The Simple Economics of Artificial Intelligence*. New York: Ingram Publisher Services, 2018.

The Economist. "The Coming Tech-Lash." November 18, 2013.

Israel, Paul. *Edison: A Life of Invention*. New York: John Wiley & Sons, 1998.

Parker, Geoffrey G., Marshall W. Van Alstyne, and Sangeet Choudary. *Platform Revolution: How Networked Markets Are Transforming the Economy and How to Make Them Work for You*. New York/London: W.W. Norton & Company, 2016.

Peneder, Michael, and Andreas Resch. "Schumpeter and Venture Finance: Radical Theorist, Broke Investor, and Enigmatic Teacher." *Industrial and Corporate Change* 24, no. 6 (December 2015): 1315–1352.

Schmidt, Eric, and Jonathan Rosenberg. *How Google Works*. New York: Grand Central Publishing, 2014.

Chapter 3

Akcigit, Ufuk, and Sina T. Ates. "What Happened to U.S. Business Dynamism?" *NBER Working Paper #25756* (May 8, 2019). https://www.nber.org/papers/w25756.

Ante, Spencer E. *Creative Capital: Georges Doriot and the Birth of Venture Capital*. Cambridge, MA: Harvard Business School Press, 2008.

Auerswald, Philip, and Lewis Branscomb. "Valleys of Death and Darwinian Seas: Financing the Invention to Innovation Transition in the United States." *Journal of Technology Transfer* 28, nos. 3–4 (February 2003): 227–239.

Barkai, Simcha. "Declining Labor and Capital Shares." *Journal of Finance* 75, no. 5 (2020): 2421–2463.

Branscomb, Lewis, and James H. Keller (eds.). *Investing in Innovation: Creating a Research and Innovation Policy That Works*. Cambridge, MA: MIT Press, 1998.

Bryan, Kevin A., and Erik Hovenkamp. "Startup Acquisitions, Error Costs, and Antitrust Policy." Symposium: Reassessing the Chicago School of Antitrust Law. *University of Chicago Law Review* 87 (2020): 331–356.

Cowen, Tyler. *Average Is Over: Powering America Beyond the Age of the Great Stagnation*. Boston: Dutton Publishing, 2013.

Cowen, Tyler. *The Great Stagnation: How America Ate All the Low-Hanging Fruit of Modern History, Got Sick, and Will (Eventually) Feel Better*. New York: Dutton Adult, 2011.

Davies, Alex. *Driven: The Race to Create the Autonomous Car*. New York: Simon & Schuster 2021.

Decker, Ryan A., John Haltiwanger, Ron S. Jarmin, and Javier Miranda. "Declining Business Dynamism: What We Know and the Way Forward." *American Economic Review* 106, no. 5 (2016): 203–207.

The Economist. "The Coming Tech-Lash." November 18, 2013.
The Economist. "Technology Firms Are Both the Friend and the Foe of Competition." Special Report, November 15, 2018.
Fujiwara, Hisanori. "What Shapes Venture Capital Firms' Expansion across the Globe? Country-Specific and Firm-Specific Factors." *Journal of Private Equity* 17, no. 1 (2013): 7–13.
Gordon, Robert. *The Rise and Fall of American Growth: The U.S. Standard of Living Since the Civil War*. Princeton, NJ: Princeton University Press, 2016.
Kaplan, David A. *The Silicon Boys: And Their Valleys of Dreams*. New York: HarperCollins, 2000.
McCraw, Thomas K. *Prophet of Innovation: Joseph Schumpeter and Creative Destruction*. Cambridge, MA: Harvard University Press, 2007.
Peneder, Michael, and Andreas Resch. "Schumpeter and Venture Finance: Radical Theorist, Broke Investor, and Enigmatic Teacher." *Industrial and Corporate Change* 24, no. 6 (December 2015): 1315–1352.
Philippon, Thomas. "The Economics and Politics of Market Concentration." *National Bureau of Economic Research*, no. 4 (December 2019). https://www.nber.org/reporter/2019number4/economics-and-politics-market-concentration.
Schumpeter, Joseph A. *Capitalism, Socialism, and Democracy*. New York: Routledge, 2010.
Schumpeter, Joseph A. *History of Economic Analysis*. New York: Routledge, 1987.
Schumpeter, Joseph A. *The Nature and Essence of Economic Theory*. New York: Routledge, 2017.
Toffler, Alvin, and Heidi Toffler. *Future Shock*. New York: Bantam Books, 1970.
Tse, Edward. *China's Disrupters: How Alibaba, Xiaomi, Tencent, and Other Companies Are Changing the Rules of Business*. New York: Penguin, 2016.
Virilio, Paul. *Polar Inertia*. San Francisco, CA: Sage Publications, 1999.

Chapter 4

Alcantara, Chris, Kevin Schaul, Gerrit De Vynck, and Reed Albergotti. "How Big Tech Got So Big: Hundreds of Acquisitions." *Washington Post*, April 21, 2021.

Baer, Ralph. *Videogames: In the Beginning*. Springfield, NJ: Rolenta Press, 2005.

Ezrachi, Ariel, and Maurice E. Stucke. *Virtual Competition: The Promise and Perils of the Algorithm-Driven Economy*. Cambridge, MA: Harvard University Press, 2016.

Gordon, Robert. *The Rise and Fall of American Growth: The U.S. Standard of Living since the Civil War*. Princeton, NJ: Princeton University Press, 2016.

Hayward, Keith. "Airbus: Twenty Years of European Collaboration." *International Affairs* 64, no. 1 (1987–88): 11–26.

Huang, Yasheng. *Capitalism with Chinese Characteristics: Entrepreneurship and the State*. Cambridge, UK: Cambridge University Press, 2008.

Huang, Yasheng. "China's Use of Big Data Might Actually Make It Less Big Brother-ish." *MIT Technology Review*, August 22, 2018.

Johnson, Keith, and Elias Groll. "The Improbable Rise of Huawei: How Did a Private Chinese Firm Come to Dominate the World's Most Important Emerging Technology." *Foreign Policy*, April 3, 2019.

Lemley, Mark, and Andrew McCreary. "Exit Strategy." *Boston University Law Review* 101 (2021): 1–102.

Malcomson, Scott. *Splinternet: How Geopolitics and Commerce Are Fragmenting the World Wide Web*. New York: OR Books, 2016.

Mayer-Schönberger, Viktor, and Yann Padova. "Regime Change? Enabling Big Data through Europe's New Data Protection Regulation." *Columbia Science & Technology Law Review* 17 (2016): 315–335.

McCraw, Thomas K. *Prophet of Innovation: Joseph Schumpeter and Creative Destruction*. Cambridge, MA: Harvard University Press, 2007.

Schumpeter, Joseph A. *History of Economic Analysis*. New York: Routledge, 1987.

Schumpeter, Joseph A. *The Nature and Essence of Economic Theory*. New York: Routledge, 2017.

Solove, Daniel J., and Paul M. Schwartz. *EU Data Protection and the GDPR*. New York: Wolters Kluwer, 2021.

Thiel, Peter, and Blake Masters. *Zero to One: Notes on Startups, or How to Build the Future*. New York: Penguin, 2014.

Vance, Ashlee. *Elon Musk: How the Billionaire CEO of SpaceX and Tesla Is Shaping Our Future*. London: Virgin Books, 2015.

Chapter 5

The Economist, "The Data Economy." Special Report, February 20, 2020.

The Economist, "Privacy in a Pandemic." April 23, 2020.

Ezrachi, Ariel, and Maurice E. Stucke. *Virtual Competition: The Promise and Perils of the Algorithm-Driven Economy*. Cambridge, MA: Harvard University Press, 2016.

International Network of Privacy Law Professionals. "A Brief History of Data Protection: How Did It All Start?" January 6, 2018. https://inplp.com/latest-news/article/a-brief-history-of-data-protection-how-did-it-all-start/.

Kelly, Makena. "Most Democrats Refuse to Back Elizabeth Warren's Big Tech Breakup Plan on the Debate Stage." The Verge, October 15, 2019.

Lemley, Mark, and Andrew McCreary. "Exit Strategy." *Boston University Law Review* 101 (2021): 1–102.

Lemley, Mark, and David McGowan. "Legal Implications of Network Economic Effects." *California Law Review* 86 (1998): 479–610.

Nahles, Andreas. "Digitaler Fortschritt durch ein Daten-für-Alle-Gesetz-Diskussionspapier der Parteivorsitzenden der Sozialdemokratischen Partei Deutschland." *SPD*, February 14, 2019. https://www.spd.de/aktuelles/daten-fuer-alle-gesetz/.

Prüfer, Jens, and Christoph Schottmüller. "Competing with Big Data." *TILEC Discussion Paper* 2017–006 (February 16, 2017).

Swire, Peter, and Yianni Lagos. "Why the Right to Data Portability Likely Reduces Consumer Welfare: Antitrust and Privacy Critique." *Maryland Law Review* 72, no. 2 (2013): 335–380.

Stucke, Maurice E., and Ariel Ezrachi. "How Pricing Bots Could Form Cartels and Make Things More Expensive." *Harvard Business Review,* October 27, 2016.

Teachout, Zephyr. *Break 'Em Up: Recovering Our Freedom from Big Ag, Big Tech, and Big Money*. New York: St. Martin's Press, 2020.

von der Leyen, Ursula. "Shaping Europe's Digital Future." *European Commission*. February 19, 2020. https://ec.europa.eu/commission/presscorner/detail/en/ac_20_260.

Chapter 6

Contreras, Jorge L., and Jerome H. Reichman. "Sharing by Design: Data and Decentralized Commons," *Science* 350 (December 2015): 1312–1314.

The Economist. "The Data Economy." Special Report, February 20, 2020.

The Economist. "Privacy in a Pandemic." April 23, 2020.

Florida, Richard. *The Rise of the Creative Class: And How It's Transforming Work, Leisure, Community, and Everyday Life*. New York: Basic Books, 2003.

Nunziato, Dawn Carla. "The Marketplace of Ideas Online." *Notre Dame Law Review* 94, no. 4 (2019): 1519–1584.

O'Neil, Cathy. *Weapons of Math Destruction*. New York: Penguin Random House, 2016.

Prüfer, Jens, and Christoph Schottmüller. "Competing with Big Data." TILEC Discussion Paper 2017-006 (February 16, 2017).

Saxenian, AnnaLee. *The New Argonauts: Regional Advantage in a Global Economy*. Cambridge, MA: Harvard University Press, 2006.

Senor, Dan, and Saul Singer. *Start-Up Nation: The Story of Israel's Economic Miracle*. New York: Twelve, 2009.

Shkabatur, Jennifer. "The Global Commons of Data." *Stanford Technology Law Review* 22 (2019): 354–411.

Stucke, Maurice E., and Ariel Ezrachi. "How Pricing Bots Could Form Cartels and Make Things More Expensive." *Harvard Business Review*, October 27, 2016.

Chapter 7

Barrington-Leigh, Christopher, and Adam Millard-Ball. "Correction: The World's User-Generated Road Map Is More Than 80% Complete." *PLoS ONE* 14(10): e0224742 (2019).

Barrington-Leigh, Christopher, and Adam Millard-Ball. "The World's User-Generated Road Map Is More Than 80% Complete." *PLoS ONE* 12(8): e0180698 (2017).

Boyle, James. *The Public Domain—Enclosing the Commons of the Mind*. New Haven, CT: Yale University Press, 2008.

Eisenberg, Rebecca S. "Patenting the Human Genome." *Emory Law Journal* 39 (1990): 721–745.

European Commission. "Creating Value through Open Data: Study on the Impact of Re-use of Public Data Resources." Digital Agenda for Europe (2015): 10.2759/328101.

Goldstein, Brett, and Lauren Dyson. *Beyond Transparency—Open Data and the Future of Civic Innovation*. San Francisco: Code for America Press, 2013.

Gurin, Joel. *Open Data Now—The Secret to Hot Startups, Smart Investing, Savvy Marketing, and Fast Innovation*. New York: McGraw Hill, 2014.

Manyika, James, Michael Chui, Diana Farrell, Steve Van Kuiken, Peter Groves, and Elizabeth Almasi Doshi. "Open Data: Unlocking Innovation and Performance with Liquid Information." McKinsey Global Institute, 2013.

Mayer-Schönberger, Viktor, and David Lazer (eds.). *Governance and Information Technology—From Electronic Government to Information Government*. Cambridge, MA: MIT Press, 2007.

Chapter 8

Allison, Graham. *Destined for War: Can America and China Escape Thucydides' Trap?* Boston: Houghton Mifflin Harcourt, 2017.

Bremmer, Ian, and Cliff Kupchan. "Risk 3: Global Tech Cold War." *Eurasia Group*, January 2, 2018.

Chinese Government Document. "Made in China 2025." Strategy Paper, July 7, 2015. http://www.cittadellascienza.it/cina/wp-content/uploads/2017/02/IoT-ONE-Made-in-China-2025.pdf.

Couldry, Nick, and Ulises Mejias. *The Costs of Connection: How Data Are Colonizing Human Life and Appropriating It for Capitalism*. Palo Alto, CA: Stanford University Press, 2019.

Couldry, Nick, and Ulises Mejias. "Data Colonialism: Rethinking Big Data's Relation to the Contemporary Subject." *Television and New Media* 20, no. 4 (2019): 336–349.

The Economist. "The Tech Cold War Is Hotting Up." July 9, 2020.

The Economist. "Technopolitics." Briefing, March 15, 2018.

Gertner, Jon. *The Idea Factory: Bell Labs and the Great Age of American Innovation*. New York: Penguin, 2012.

Gray, Mary, and Siddharth Suri. *Ghost Work: How to Stop Silicon Valley from Building a New Global Underclass*. Boston: Houghton Mifflin Harcourt, 2019.

Hope, Janet. *Biobazaar—The Open Source Revolution and Biotechnology*. Cambridge, MA: Harvard University Press, 2008.

Institute for Global Dialogue. "The G20 Osaka Summit: Japan's Delicate Balancing of Diplomacy-Led and Development-Led Engagements." Proceedings Report (2019): 1–13.

Johnson, Steven. *How We Got to Now: Six Innovations That Made the Modern World*. New York: Riverhead Books, 2014.

Khanna, Parag. "Pillar or Pawn." *Rest of the World.org*, December 11, 2020. https://restofworld.org/2020/pillar-or-pawn/.

Lee, Kai-Fu. *AI Superpowers: China, Silicon Valley, and the New World Order*. Boston: Mariner Books, 2018.

Milner, Greg. *Pinpoint: How GPS Is Changing Technology, Culture, and Our Minds*. New York: Norton, 2017.

Riordan, Michael, and Lillian Hoddeson. *Crystal Fire: The Birth of the Information Age*. New York: Norton, 1997.

Tse, Edward. *China's Disrupters: How Alibaba, Xiaomi, Tencent, and Other Companies Are Changing the Rules of Business*. New York: Penguin, 2016.

Watzinger, Martin, Thomas A. Fackler, Markus Nagler, and Monika Schnitzer. "How Antitrust Enforcement Can Spur Innovation: Bell Labs and the 1956 Consent Decree." *American Economic Journal* 12, no. 4 (2020): 328–359.

Woodcock, Jamie, and Mark Graham. *The Gig Economy; A Critical Introduction*. Cambridge, UK: Polity, 2020.

Index

Abe, Shinzo, 163
acquisition of start-ups: Google acquisition of ITA, mandated data access for, 109–111, 118, 123; public offerings of companies (IPOs), 58, 60, 83. *See also* "kill zone" (buyouts of start-ups)
Adams, Abigail, 152
Africa: clickworkers in, 153; data colonialism in, 159, 161–162; privacy regulations and data access mandates, 98
AI (artificial intelligence): DeepMind (Google), 21, 90; humans doing the work in the background, 151–152; machines replacing human workers, 151–152; market concentration as accelerated by, 88; structural advantage of China in battle for supremacy, 156. *See also* machine learning
Airbus, 72–73, 74
algorithms: academic researchers as source of, 33; claimed to be an issue in success of Big Tech, 31, 33–34, 35; as open-source, 33–34; retail recommendations, 22
Alibaba: antitrust crackdown, 9–10, 66, 87; cloud services, 73, 74; emergence of, 63; Freshippo, 63–64; government access to data of, 65; recommendation algorithms of, 22. *See also* narrative of achievement
Alipay, 10, 27, 66
Allison, Graham, 157
AlphaGo, 90
Altmaier, Peter, 73
Amazon: acquisition of start-ups ("kill zone"), 58; Alexa, 50; breakup of, calls for, 80–81; and COVID-19 pandemic, 16; defensive battle against data access, 14; as hyperscaler, 73; lobbying against mandated data access, 14; in mandated data access examples, 111, 112; market concentration of, 47; Marketplace, 26; Mechanical Turk, 151, 152; Prime service, 47; recommendation algorithms of, 22;

Amazon *(continued)*
reuse of data, 121; venture capital and, 45; Whole Foods, 81. *See also* Big Tech; narrative of achievement
Amazon Web Services (AWS), 31–32, 73
American Research and Development Corporation (ARDC), 44, 48–49
American Revolution, 2, 130, 168
anonymized data. *See* nonpersonal data
Ant Financial, 27
Ant Group (China), 10, 66
anti-capitalist intellectuals, 40, 46
antitrust actions: and Alibaba (China), 9–10, 66, 87; AT&T, 81; AT&T and Bell Labs consent decree (1956), 166–168; breakups of Big Tech companies, 80–81, 87; calls for, 80–86; Chinese policy not to pursue, 66–67; European Union and, 9, 67–69, 86–87; existing law as ill-equipped to deal with Big Tech, 84, 85; Facebook (US), 9, 82; Google (US and Europe), 9, 68–69, 82; incremental measures proposed to prevent "kill zone" acquisitions, 83–84, 86, 87; Microsoft (US), 67, 83; policy choice of US government authorities not to pursue, 51–52, 81–83, 109; policy of Biden administration to pursue, 83, 84–86. *See also* monopolies; regulation of Big Tech
Apollo (autonomous vehicle), 20–21, 22
Apollo Go, 21
Apple: acquisition of start-ups ("kill zone"), 58, 59–60; Apple News, 30; autonomous vehicle, 20, 121; as breaking with Silicon Valley's culture of free flow of information, 108–109; buying up high-tech music companies, 60; Chinese regulation of, 113; COVID-19 and refusal to help with smartphone tracing apps, 4–5; COVID-19 pandemic and mobility data on lockdowns, 16, 146; defensive battle against data access, 14; geographical location data, 121; iPhones, 63, 121; lobbying against mandated data access, 14; Mac, 50, 107–108; relocation of facilities from China, 156; reuse of data, 121; and surveillance capitalism, 70. *See also* Big Tech; narrative of achievement; smartphone app stores
Arab Spring, 7
Asia: clickworkers in, 153; cloud service providers and, 32; COVID-19 pandemic and tracking infections, 3; data colonialism in, 159, 161–162; privacy regulations and data access mandates, 98; talent drain to Big Tech, 98; and the "Tech Cold War" (China and the US), 157
AT&T: antitrust breakup of, 81; Bell Labs and the data access mandate (1956), 166–168; as monopolist, 165; regulation of, 84–85

Atari, 56
Australia: open science research and, 126; privacy regulations and data access mandates, 98; smartphone pandemic tracing apps crippled in, 4–5; social media licensing requirements for local news content, 113
authoritarian governments: China's political authoritarianism, 65, 70; erosion of trust in government and rise of, 129–130
automakers, 27–28
autonomous vehicles, 18–21, 22, 27–28, 122
AutoX, 20
AWS (Amazon Web Services), 31–32, 73
Azure, 32, 73, 96

Baer, Ralph Henry, 55
Baidu: autonomous vehicles, 20–21, 22; and centralization of power, 8; emergence of, 63; government access to data of, 65; iQiyi, 63. *See also* narrative of achievement
Bangladesh, clickworkers in, 153
banking, 23, 27, 48, 90; venture debt (loans), 83
Barcelona, 15
Barkai, Simcha, 53
Beats headphones, 60
Bell Labs: data access mandate by US government (1956), 166–168; inventions of, 44, 166, 167
Berners-Lee, Tim, 7, 50, 134–135
Bezos, Jeff, 8
Bezos' Law, 31–32
Biden administration, 9, 66, 83, 84–86, 156
Big Bang Theory (TV), 34, 36
Big Tech: as breaking with the collaborative innovation culture of Silicon Valley, 107–109, 125; GAFAs (Google, Amazon, Facebook, Apple), 113; hype cycle, 64; lobbying against data access mandates, 14; open data, voluntary donations of, 146; political polarization and radicalization as fueled by, 148, 164; as projecting American global tech dominance, 82; reuse of data by, 121, 123; smartphone apps for tracing COVID-19 infections blocked by, 3–5; structural advantages of, 87–88, 123; and surveillance capitalism, 8, 70; vertically integrated organizational structures of, 125. *See also* algorithms; China; data access mandates; data colonialism; exclusive access to data by Big Tech; informational power asymmetries; "kill zone" (buyouts of start-ups); market concentration; monopolies; narrative of achievement; regulation of Big Tech; Silicon Valley; "Tech Cold War" (China and the US); *specific companies*
Bing Travel, 110
biogenetics, 122
biotech companies, open science research and, 126, 127–128
Bismarck, Otto von, 157

Boeing, 72, 73
Booking.com, 26, 30, 35
Boston, Route 128, 105–107, 108, 109, 125
Bradford, Andrew, 2
Branscomb, Lewis, Darwinian Sea, 75–79, 97
Brazil, 153
Bremmer, Ian, 154–155
Breton, Thierry, 69, 72
Bria, Francesca, 15
Brin, Sergey, 8
British colonialism, and the US, 1–2, 168. *See also* colonialism; data colonialism
Broadway shows, investors, 44
bureaucracies: as adversarial to open data, 139; confidentiality principle as bar to open data, 138, 140–141; and information asymmetries, 141
Bush, George W., 56, 82, 83
buyouts. *See* "kill zone" (buyouts of start-ups)
ByteDance, 35

California, noncompete clauses not enforced in, 107
California Consumer Privacy Act, 71
California Data Collaborative, 117
California Department of Motor Vehicles, 19
Cambridge Analytica, 80
cancer research and treatments, 22, 27, 128
capital investment: "founder's profit," 45; importance to start-ups, 41; postwar era and capital investment, 42, 44–45. *See also* venture capital
capitalism: change as the essence of (Schumpeter), 38–39; classical economists' concern with equilibria in, 38, 40; clickworkers and division of labor in, 151; surveillance capitalism, 8, 70. *See also* Big Tech; creative destruction of entrepreneurial innovation; digital laggards; monopolies; venture capital
Carnegie Mellon University, 19
Castells, Manuel, 6
CDC (Centers for Disease Control), 136
Celera, 127–128
Central America, clickworkers in, 153
centralization. *See* Big Tech; informational power asymmetries
Cerf, Vint, 7
CERN (European Organization for Nuclear Research), 134–135
CheapTickets, 110
Chime, 27
China: antitrust actions against Big Tech taken by, 9–10, 66, 87; autonomous vehicles, 20–21; capitalism with Chinese characteristics, 63–67, 76–77, 154–155; and centralization of economic power, 8; cloud service providers, 32; and competition among start-ups, 64–65; and COVID-19 pandemic, 126, 127; as data colonialist power, 153, 159; data security questions, 156; and the hype cycle, 64; innovation pace as

slowing in, 76–77; open science research and, 126; political authoritarianism and, 65, 70; private capital and, 65; procurement contracts by government, 65; protectionism against foreign competition, 63, 156; protectionism of the US against, 66, 158; regulation of Big Tech, 113, 156; state-owned companies (SOEs), 64; superiority of products of, 63–64, 154–155; talent drain to the US, 98; talent joining Big Tech of, 98; in Western markets, 156–157. *See also* Big Tech; data colonialism; "Tech Cold War" (China and the US)
Christensen, Clayton, 28
clickwork: and capitalist division of labor, 151; data colonialism and, 13, 153, 159; dependence of laborers on, 152; human prep work behind AI, 150–152; locations of workers, 153; machines replacing humans in, 151–152, 159; pay rates and working conditions, 152–153, 159; treated as independent contractors/self-employed, 152; types of work done, 151
Clickworker platform, 151
Clinton, Bill, 128, 137, 139
cloud computing: China and, 32; as commodified product, 33, 74, 78; decreasing cost of, 31–33, 74; European GAIA-X system, 32, 72–75; increasing capacity of, 31, 74
Coase, Ronald, 96
Cognizant, 151
Cohen, Julie, 8
Cold War: and Bell Labs data access mandate, 166, 167–168; capacity for innovation as competitive advantage in, 42; division of the world into two camps, 155; end of, and Pax Americana, 154; skirmishes between surrogate forces, 155. *See also* "Tech Cold War" (China and the US)
Coleco, 56
collective action, problem of, 120
colonialism: British, in the US, 1–2, 168; data colonialism as turning the tables on former colonial powers, 13, 153, 157, 159; decolonization and postcolonial governance, 161, 162–163; self-image of Indigenous peoples and, 161; suffering caused by, 160. *See also* data colonialism
commodities, cloud services as, 33, 74, 78
Common Voice project, 145
competition: buying out threats (kill zone), 47–48, 59–61, 62, 85; capacity for innovation as advantage in, 42; capacity for innovation declines, and decrease in, 51–52; Chinese start-ups and, 64–65; and commodity pricing, 33, 74, 78; data protection laws marketed as advantage in, 71–72; EU Commission and importance of, 68; exclusive access to data as advantage in, 53, 92; lack of, and rising prices/profit margins, 51, 52, 53, 65, 77, 91, 97; open data as

competition *(continued)*
reinvigorating, 147; protectionism as not having built Silicon Valley, 76; protectionism by China, 63, 156; protectionism by the US, 66, 158; Silicon Valley and "coopetition," 107–109; Peter Thiel on ("for losers"), 61, 62. *See also* antitrust actions; creative destruction of entrepreneurial innovation; "kill zone" (buyouts of start-ups); market concentration; monopolies; Schumpeterian competition
Compton, Karl, 44
computing power, claimed to be issue in success of Big Tech, 31–33, 34–35
Conference on Trade and Development (United Nations, 2019), 162–163
confidentiality: and data not subject to access mandates, 111; principle of, as barring open government data, 138, 140–141. *See also* data protection, privacy laws (minimal data use)
Constitution (US), 2–3
convenience, and user willingness to provide information, 10, 26, 159
Cook, Tim, 4–5, 8
Cooper, Dr. Sheldon (*Big Bang Theory* TV), 34, 36
coopetition, 108
copyrights, 130
Couldry, Nick, 153
COVID-19 pandemic: data colonialism and control of information about, 159; mobility data for assessing effects of partial lockdowns, Big Tech providing, 16, 146; open data initiatives and, 15–16, 126–127, 133, 136, 146; political rifts and, 127; and privacy laws (minimal data use), 4–5, 100–101; regional mobility data and, 3; smartphone apps for tracing infections, Big Tech blocking, 3–5; tracking infections manually, 3
Cowen, Tyler, 51
C programming language, 167
creative destruction of entrepreneurial innovation (Schumpeter): the aging of Big Tech companies and lack of, 57; capital investment and, 41; definition of, 38–39; market concentration and lack of, 47, 52, 62–63, 85; monopolies as swept aside by (Schumpeter), 40–41, 52; Elon Musk and, 165; the narrative of achievement as protecting Big Tech from, 36; Schumpeter's nightmare that lack of resources for entrepreneurs would spell the end of, 41–42, 46, 79, 87–88; threat of, and the "kill zone" of buying out start-ups, 47–48, 60–61; venture capital as thwarting, 60, 85; video game consoles as illustration of, 55–56. *See also* innovation, capacity for; "kill zone" (buyouts of start-ups); Schumpeter, Joseph Alois
credit card fraud detection systems, 22–23
credit scoring systems, 23

CrowdFlower, 150
Cruise (autonomous vehicle), 20
cystic fibrosis, 128

Darwinian Sea of innovation, 75–79, 97
data: air as metaphor for, 104; as common good, 15; democratization of, 15; digital laggards and functional approach to, 23–24; as digital publicly accessible good, 14; exclusion right to, 95–96; no legal ownership of, 14, 95–96; as "non-rivalrous good," 14, 91–92; as not like oil, 23–25, 28, 92–93; oil as parallel to exclusive access, 35; as "property," as little effect against Big Tech, 121; as public good, 127, 130; reuse of, 121–123; unused, as destroying value, 93, 103; unused, as hostile to innovation, 103; unused, statistics on, 15, 93, 97; use as basis of the value of, 14, 17, 24–25, 28, 93, 124. *See also* data access mandates; data colonialism; exclusive access to data by Big Tech; open data
data access mandates: overview, 11–17, 79; Big Tech lobbying against, 14; competition for best use of the same data, 92–93, 103, 143; continued innovations of companies giving access, 167; and data colonialism, end of, 168; data literacy as mutually reinforcing with, 125; and digital laggards, transformation of, 93; diverse data sets, 94, 112, 119–120; the economy as widely stimulated by, 88, 89–91, 167; emerging economies and, 13, 163; and empowerment of the many, 93, 102, 103, 104, 168; enforcement, 113–114; the EU and nations considering, 11, 89–90, 102, 110–111; expropriation fears, 14, 92–93; as global opportunity for new Silicon Valleys, 109, 125; incentives for collection vs. use of data, 93–94, 97, 103; increased data access for all, 103; innovation as widely stimulated by, 88, 97, 121–125, 167; mindset toward data as critical factor in, 123–125; multiple purposes for the same data, 92; and participation in an ecosystem of innovation, 114; and privacy laws, modification of principle of, 14, 98, 102; as pro-active vs. defensive strategy, 11, 123; as pro-market and pro-competition strategy, 88–89, 91, 92–93; proposals for organizing, and digital decolonization, 162–164; small and medium-sized companies, advantages for, 112, 119–120, 167; and the "Tech Cold War" (China and the US), 11–12; and unused data as issue, 15, 93, 97, 103; value as created by use of data, 14, 17, 24–25, 28, 93, 124. *See also* data; data access mandates, historical; data access mandates, policies; open data
data access mandates, historical: Bell Labs, 166–168; earliest data

data access mandates *(continued)*
protection laws, 98–99; Google acquisition of ITA (Google Flights), 109–111, 118, 123; Sweden and long tradition of, 99. *See also* data access mandates; open data initiatives

data access mandates, policies: categories of data, division into, 117–119; confidential data not subject to, 111; data protection used with, 111–112; decentralized vs. centralized storage and transmission of data between data requesters and -holders, 114–116, 118, 119; enforcement, 113–114, 119; government involvement with transmission not needed, 116, 118, 119; government providing online directory of data and enforcement, 119; liability rules, 120; limits on number of annual requests permitted, 118; metadata requirements, 119; nonpersonal/depersonalized data as included in, 111–112, 120; online directory for data access, 118, 119; open government data requiring private sector mandates as adjunct to, 142–143; open standards for, 119–120; open- vs. closed system of data access, 116–117; request process, 117–119; requests for data open to all, with reciprocal sharing, 112–113; reuse of data, 121–123; security risks, and "data fiduciaries" concept, 115–116, 118; size threshold for companies mandated to share, 112; size threshold, determination of, 113; size threshold, and percentage of data required to share, 112, 113, 118; small and medium-size companies, measures to assist, 119–120; small businesses not required to share data, 112; technical implementation not to be specified by law, 119; term limits/sunset clauses cannot be included in, 111, 123; time requirements to answer requests, 118; trade secrets not subject to, 111; two-way sharing, 103–104, 112–113

data alchemy, 21–23, 34

data colonialism: overview, 13; appropriation of data made to look natural, 160; clickworkers and, 13, 153, 159; data access mandates and the end of, 168; and decolonization/postcolonial governance of political colonialism, 161; decolonization proposals, 162–164; definition of, 153; democratic decision-making controlled by, 4–5; former European colonial powers and former colonialized nations as both ruled by, 13, 153, 157, 159; informational power asymmetries and, 13, 159; as metaphor, 157, 160; and the narrative of achievement (Big Tech), 160–161; policies of digital sovereignty in resistance to, 161–162; rejection of colonialist proposals, 162; and the "Tech Cold War" (China and the US), 159; technical structures of dominance, 160; United States and China as the

colonial powers in, 153, 159; users exporting their data to Big Tech and, 153–154, 159. *See also* "Tech Cold War" (China and the US)
Data for Good (Facebook), 146
Data General, 105, 106
data literacy: as mindset, 125; and open government data, 141
data markets, as failure, 94–97, 103–104
data protection laws: and abuse potential of empowerment, 148–149; data access as original aim of, 98–99, 101; data access regulations and, as two sides of the same coin, 14, 99; data portability right, 101, 120; as defensive strategy, 10; and democracy, promotion of participation in, 70–71, 99; digital laggards as further disincentivized by, 11, 102; digital start-ups as hampered by, 11; empowerment as original aim of, 10–11, 98–99, 101; empowerment in new paradigm of, 102; information asymmetries and centralization growing under, 4–5, 10–11, 14, 101–102, 148–149; information asymmetries meant to be dismantled and prevented by, 99, 102; intentions of, 10, 70–71, 98–99, 102; mandated data access and modification of, 14, 98, 102; mandated data access contained in earliest forms of, 98–99; marketed as competitive advantage, 71–72; narrow definition of restrictions imposed by, 130; open data and continued need for, 130, 148–149; paradigm shift in, 102; privacy protections as goal of, 70–71; privacy protections becoming narrow focus of, 99–100, 102; responsibility for, as lying with those who derive the most economic benefit, 148–149. *See also* data access mandates; data protection, privacy laws (minimal data use); General Data Protection Regulation (GDPR)
data protection, privacy laws (minimal data use): and artificial scarcity of data, 102; consent as issue in, 10, 101–102, 148; COVID-19 pandemic and weaknesses of, 4–5, 100–101; data access mandates and modification of principle of, 14, 98, 102; dysfunction as the price of, 101; and economic indicators, slowing of, 71–72, 99–100; Google/Apple argument against smartphone COVID-19 tracing apps based on, 4–5; legal boundary between meaningful and harmful use, 99; as overburdening individuals in managing privacy settings, 10–11, 71, 148; as quasi-religious dogma, 100, 102. *See also* data protection laws; individuals exporting their personal data to digital platforms; nonpersonal data
data standards. *See* standards
data use regulation. *See* data access mandates
dating platforms, 47

decentralization. *See* data access mandates; open data
Decker, Ryan, 51
DeepMind, 21, 90
Defense Advanced Research Projects Agency (DARPA), 18–19
delivery portals, 26, 47
Dell, relocation of facilities from China, 156
democracy: data protection laws promoting participation in, 70–71, 99; exclusive access to data by Big Tech as threat to, 130, 149; and the free press/open access to information, 1–3, 5, 8, 130, 168; open data and health of, 129–130, 138–139, 147–148, 168; oversight of, 148; and the right to access information, 130–131. *See also* democratic flow of information
democratic flow of information: and betrayal of the promise of the digital revolution, 7–8; as counter to Big Tech, 5–6, 99; Google/Apple exercising veto over, 4–5
depersonalization of data, 111–112, 120, 140. *See also* nonpersonal data
Didi Chuxing, 63
Digital Equipment Corporation (DEC), 44–45, 105, 106
digital laggards: and access to public transport timetables, 143; "data as the new oil," 24, 28; data protection laws as further disincentive to participation, 11, 102; "digital transformation" as functional approach to data, 23–24; dilemma of path dependency for, and difficulty of changing business models, 28; failure to invest in the digital marketplace, 26–28; fear of sharing their own data, 103–104; mandated data access and transformation of, 93; open data and transformation of, 147; as unable to reuse their own data, 121; young and data-rich Silicon Valley companies as causing increasing difficulties for, 90. *See also* small and medium-sized companies, and mandated data access
digital marketplace: convenience of, 26; digital dividend of, as unevenly distributed, 91; fees and, 26; investment in, vs. digital laggards who did not, 26–28; as no longer competitive, 91; open data as lowering barriers to, 147
digital payment market, 27, 63, 66
digital sovereignty: and EU industrial policy for GAIA-X cloud provider, 72, 75; personal, as overburdening individuals, 10–11, 71, 148; and policies of resistance to data colonialism, 161–162
Disney, 27, 47
DJI, 63
DoorDash, 26
Doriot, Georges, 42–45, 50, 105
drones, 63
drugs and drug companies, 16, 51; and access to Human Genome Project data, 128

earn-out agreements, 83–84
Eastern Europe: and Gazprom

market power, 68; planned economies of, 46
eBay, 113
economic inequality, increases in, 164
Edison, Thomas, 36
education, and open data, 147–148
Eisenhower administration, 167–168
Electronic Freedom of Information Act (E-FOIA), 139
emerging economies: data colonialism in, 13, 162, 163; and mandated access to data, 13, 163; rejection of the "Osaka Track" data cooperation proposal, 163. *See also* data colonialism
empowerment: data access mandates and, 93, 102, 103, 104, 168; data colonialism betraying the promise of, 153; informational power asymmetries betraying the promise of, 7–8; open data and, 130, 149; as original aim of data protection laws, 10–11, 98–99, 101; paradigm shift in data policy toward broad empowerment, 102; voluntary open access and, 15
Engelbart, Douglas, 50
entrepreneurs: as the microeconomic center, 39. *See also* creative destruction of entrepreneurial innovation
environmental data and sustainability, 145, 147, 164–165
ePrivacy Directive, 100
Ericsson, 63
EuroMOMO (European Mortality Monitoring Project), 136
Europe: Marshall Plan, 42. *See also* colonialism
European Court of Justice, 9, 68, 69
European Nucleotide Archive, 128
European Parliament, and antitrust actions, 80, 86–87
European Union: Airbus, 72–73, 74; antitrust actions against Big Tech by, 9, 67–69, 86–87; competition Commissioner, importance of, 69; data colonialism in, 153, 159, 161–162, 163; and democracy, promotion of political participation in, 70–71; digital sovereignty, quest for, 10–11, 71, 72, 75; economic decline and privacy laws, 71, 71–72, 99–100; "European values," 70; industrial-age policy and GAIA-X cloud provider, 32, 69, 72–75; "kill zone" acquisition of start-ups from, 90, 99; as lagging behind in innovation, 71–72; mandated access to data being considered by, 11, 89–90, 102, 110–111; open science research and, 133; and open- vs. closed structures for mandated access, 117; policy-making through symbolism, 74–75; and privacy, protection of, 70–71; smartphone pandemic tracing apps crippled in, 4–5; surveillance state, fear of, 69–70, 72; talent drain to Big Tech, 97–98; and the "Tech Cold War" (China and the US), 156, 157; and "unicorn" start-ups, hope for, 89–90. *See also* Big Tech; data colonialism; data protection laws; digital laggards; General Data Protection Regulation (GDPR); "Tech Cold War" (China and the US)

Eurostat, 136
evolution, and innovation (Darwinian Sea), 75–79, 97
exclusive access to data by Big Tech: appearing to be the natural order of things, 160; as break from Silicon Valley's cultural history of free flow of information, 108–109, 125; as business model, 25–28, 36; centralization of power and, 8, 149; as competitive advantage, 53, 92; and decreased capacity for innovation, 53, 87–88, 125; and hiring of talent, 53; informational power asymmetries protected by, 30–31, 35–36; lobbying to protect, 14; and the narrative of achievement, 30–31, 35–36; as structural advantage, 87–88; as threat to democracy, 130, 149. *See also* data access mandates; informational power asymmetries

Facebook: acquisition of start-ups ("kill zone"), 58, 59; advertising revenues, 47; antitrust actions against, 9, 82; Australian law requiring licensing for local news content, 113; as breaking with Silicon Valley's culture of free flow of information, 108–109; breakup of, calls for, 80–81; buying out potential competitors, 59; COVID-19 pandemic and open data initiative of ("Data for Good"), 16, 146; defensive battle against data access, 14; growing unease with power asymmetries of, 8–9; India rejecting data colonialist proposal from, 162; Instagram, 47, 48, 81, 82, 85; lobbying against mandated data access, 14; market concentration of, 47; political polarization and radicalization fueled by, 148, 164; and the regulatory tidal wave of 2019, 80; and surveillance capitalism, 70; Peter Thiel and, 61; WhatsApp, 47, 48, 59, 81, 82. *See also* Big Tech; narrative of achievement
facial recognition software, 150–151
fact checking, 136
FAIR (findable, accessible, interoperable, and reusable), 133
Fairchild Semiconductors, 44, 45, 56
Farecast, 48
Figure Eight, 151
filter bubbles, 8
financial services, 51, 137
financial technology (fintech) companies, 27
Firefox browser, 145
"flat world," 153, 164
flight bookings, and access mandate, 109–111
Flipkart, 22
forests and forestry, 145
Fowler, Dr. Amy (*Big Bang Theory* TV), 34, 36
France: *Centre de Perfectionnement aux Affaires,* 43; GAIA-X cloud provider and, 73; INSEAD business school, 43; regulation of Big Tech, 113
Franklin, Benjamin, and open access to information, 1–3, 5, 8, 130, 168

freedom of information rules, 139
Freshippo, 63–64
Friedman, Milton, 39
Friedman, Thomas, 153

G8, open science research support by, 133
G20: open science research support by, 133; "Osaka Track" data cooperation, 163
GAIA-X cloud provider, 32, 72–75
gaming and gambling platforms, 47
Gates, Bill, 47
Gazprom, 68
General Data Protection Regulation (GDPR): overview, 10, 70–71; COVID-19 pandemic as illustration of problems with, 3–5, 100–101; data portability right, 101, 120; economic decline and, 71–72; enforcement of, 101, 113; as facilitating the power of Big Tech, 10–11, 101–102; paradigm shift in, toward empowerment, 102; personal sovereignty over data, as overburdening individuals, 10–11, 71; privacy activists wishing to emulate, 10, 11, 71; privacy laws becoming the narrow focus of, 99–100, 102; trust established by, 114. *See also* data protection laws
General Electric, 27, 112
Genomic Data Commons (USNIH), 128
geophysics, 129
Germany: autonomous vehicles, 20; and data access mandates, 98–99, 110; GAIA-X cloud provider and, 73; and World War I, 157; and World War II, 41–42, 43, 44, 55
GitHub, 33–34, 60, 116–117
globalization of Big Tech, lack of regulations on, 154
global regions: converging to form a data access coalition, 163–164; opportunity to become like Silicon Valley, 109, 125
Global South: cloud service providers and, 32; and data decolonization, 163; and the "Tech Cold War" (China and the US), 157. *See also* data colonialism
Google: acquisition of start-ups ("kill zone"), 58, 59, 90; advertising revenues, 47; AlphaGo, 90; antitrust actions against, 9, 68–69, 82; as breaking with Silicon Valley's culture of free flow of information, 108–109; breakup of, calls for, 80–81; COVID-19 and refusal to help with smartphone tracing apps, 4–5; COVID-19 pandemic and mobility data on lockdowns, 16, 146; DeepMind (machine learning) subsidiary, 21, 90; defensive battle against data access, 14; and informational power, 6–7; ITA acquisition (Google Flights), 109–111, 118, 123; lobbying against mandated data access, 14; in mandated data access examples, 111, 112; market concentration of, 47; NEST thermostat, 121; Public Data Explorer, 136; reuse of data, 121; Street View cars, 27–28; and surveillance capitalism, 70;

Google *(continued)*
venture capital and, 45; voice recognition software and, 27; Waymo subsidiary, 19, 20–21, 22, 151; Waze acquisition, 59, 81, 85; YouTube, 47, 48, 113. *See also* Big Tech; narrative of achievement; smartphone app stores
Google Maps, 27–28, 59, 144–145
Google Now, 50, 90
Gordon, Robert, 51
governments: erosion of trust in, and rise of authoritarian rule, 129–130; and mandated data access, government providing online directory of data and enforcement, 119; and mandated data access, not needed for involvement in transmission of, 116, 118, 119; open science requirement for funding by, 132, 133; open science research facilitated by, 127–128; postwar era and capital investments, 42; right to access information (UN International Covenant on Civil and Political Rights), 130–131; as supporting specific technologies vs. supporting innovation, 64–65, 74–79. *See also* authoritarian governments; bureaucracies; democracy; industrial-age policy-making; open government data; regulation of Big Tech
GPS, 137–138
Graham, Mark, 152
Gray, Mary, 150–151, 152
Great Depression, 39
Greeks, ancient, 157

hackers, and centralized storage and transmission for mandated data access, 115–116
Hara, Kotaro, 152
health care: lack of competition in, 51; open data and, 147–148; reuse of data and, 122; Silicon Valley startups causing difficulties for digital laggards in, 90. *See also* COVID-19 pandemic; drugs and drug companies; medicine
hepatitis, 128
Hesse, Germany, 98–99
historical access. *See* data access mandates, historical
Hoefler, Don, 106
Hoffman, Reid, 78
Hollywood studios, 27
HP, relocation of facilities from China, 156
Huang, Yasheng, 63
Huawei: government procurement contracts and, 65; and the "Tech Cold War" (China and the US), 154, 155, 157; technological superiority of products, 63, 154–155
Human Genome Project (HGP), 127–128
human resource management, 122
Humvee, 19
Hyundai, 28

IBM, 27
independent contractors, clickworkers treated as, 152
India: clickworkers in, 153; and data access mandates, 163; and data

colonialism, 162, 163; as emerging economy, 13; rejection of Facebook's data colonialist proposal, 162; rejection of the "Osaka Track" data cooperation proposal, 163; talent drain to Big Tech, 98. *See also* data colonialism
individuals exporting their personal data to digital platforms: convenience and, 10, 26, 159; data colonialism and, 153–154, 159; data protection laws as overburdening the individual, 10–11, 71, 148; to foreign Big Tech while domestic start-ups are hampered by data protection laws, 11. *See also* data protection
Indonesia, rejection of the "Osaka Track," 163
industrial-age policy-making: and data colonialism, 162; and EU's GAIA-X cloud provider project, 32, 69, 72–75; vs. governmental support for innovation, 64–65, 74–79. *See also* industrialization
industrialization: East Coast (Route 128) tech companies and business model of, 107; and rethinking the production process, 124; waves of, 124
"informationalism," 6
informational power asymmetries: Big Tech and establishment of, 6–7; business model and, 26; capital markets and, 41; data access as counter to, 5–6; data colonialism and, 13, 159; data protection laws and goal of countering, 99, 102; data protection laws and growth of, 4–5, 10–11, 14, 101–102, 148–149; and decrease in capacity for innovation, 53; end of, and new era of data sharing, 16–17; exclusive access to data and protection of, 30–31, 35–36; growing unease with, 8–9; lack of engagement with question of, 6; machine learning and, 22; media power, 8; the narrative of achievement of Big Tech and, 30–31, 35–36; and power, definition of, 6; the promise of political and social empowerment betrayed by, 7–8; proprietary data standards and, 120; Schumpeter's duel in protest of, 53–54; suppliers at the mercy of, 26, 30; Sweden's historic data access laws to counter, 99; technical infrastructure and, 25. *See also* exclusive access to data by Big Tech
information technologies (IT): shift from industrial- to knowledge-based societies, 49; venture capital as enabling all the fundamental shifts in, 45–46, 49, 85
initial public offerings of companies (IPOs), 58, 60, 83
innovation: collaborative innovation as culture of Silicon Valley, 107–109, 125; Darwinian Sea of, 75–79, 97; follow-on innovation after mandated data access, 167; open data as prerequisite for

innovation *(continued)*
rekindling the fire of, 149; venture capital as, 48–49. *See also* creative destruction of entrepreneurial innovation; innovation, capacity for; Schumpeterian innovation
innovation, capacity for: access to data as necessary prerequisite for, 53; Chinese start-ups and, 64–65; as decisive competitive advantage in the Cold War, 42; as decreasing in the Here and Now, 49–53, 164; EU as lagging behind in, 71–72; governments supporting, vs. supporting specific technologies with industrial-age policies, 64–65, 74–79; informational power asymmetries and decrease in, 53; market concentration and decrease in, 51–52; measurement of, 49; monopolies as slowing the pace of, 76–77, 79, 85; postwar era and access to capital, 42–43; Schumpeter on the importance of entrepreneurs to, 78; Schumpeter's nightmare of loss of, 41–42, 46, 79, 87–88, 97, 123; social division as fostered by decreases in, 164–165; societal and community support for, 78–79; venture capital facilitating, 45–46, 49, 85
Instagram, 47, 48, 81, 82, 85
INTEL, 45
Intellivision, 56
International Covenant on Civil and Political Rights (United Nations), 130–131
Internet: data sources, resource for choosing, 136; free flow of information enhanced by, 108; GPS and, 137; open science research and, 126, 134, 136. *See also* World Wide Web
Internet Explorer, 67
iQiyi, 63
Israel, and innovation, 109
ITA Software, 109–111, 118, 123

Japan: autonomous vehicles, 20; data colonialism in, 153, 159, 163; "Osaka Track" international data coordination proposed by, 163; and the "Tech Cold War" (China and the US), 157
jobs and freelancing platforms, 47
Johns Hopkins University, 136
Johnson, Boris, 4

Kayak.com, 110
Keimer, Samuel, 1
Keynes, John Maynard, 39, 40
Khan, Lina, 9, 84–85
Khanna, Parag, 157
"kill zone" (buyouts of start-ups): calls for regulation of, 83–84, 85–86; competitive threat to Big Tech ended via, 47–48, 59–61, 62, 85; definition of, 47–48; and European start-ups, 90, 99; EU rules to prevent buyouts of European start-ups, 87; founders planning to "cash out" as exit strategy, 57, 58, 60, 86, 88; founders who refuse to sell, 48; of "grown-ups," 60; incremental

regulations proposed to prevent, 83–84, 86, 87, 123; mandated data access as mediating, 109–111, 118, 123; numbers of, 58; premium prices paid for, 59, 60; profitability of, as political poison for regulators, 86; public offering of companies (IPOs) compared to, 58, 60, 83; talent acquired via, 58–59; venture capital and, 48, 58, 60, 86. *See also* Big Tech
Klarna, 27
Kleiner, Eugene, 45
Kleiner Perkins, 45
Korean airliner, shot down by Soviets, 137

Latin America. *See* Central America; South America
lawyers, legal bots, 23
League of Nations, 162–163
Lemley, Mark, 56–58, 60
Lenovo, 154–155
liability, and mandated data access, 120
life sciences, open research data and, 128
LinkedIn, 47, 60, 78
Li, Robin, and centralization of power, 8
logistics, transportation, 122, 137–138
Lyft, autonomous vehicle, 20

Mac, 50, 107–108
McCreary, Andrew, 56–58, 60
machine learning: applications of data alchemy, 22–23; autonomous vehicles and, 20–21, 22; clickworkers training their machine replacements, 151–152; DeepMind (Google subsidiary), 21, 90; informational power asymmetries and, 22; open-source algorithms for, 33–34; and partial automation of the process of innovation, 21–22; as similar to human learning, 23. *See also* AI (artificial intelligence)
McKinsey Global Institute, 138
McNamee, Roger, 8–9
Macron, Emmanuel, 4
Magnavox Odyssey, 56
Ma, Jack, 9–10
mandates. *See* data access mandates
manufacturing, 90, 124; smart factories, 23, 122. *See also* industrialization
Manyika, James, 138
Ma, Pony, 8
market concentration: Chinese policies for, 65–66; and decreasing capacity for innovation, 51–52; as historically unprecedented, 40; and the "kill zone," 60–61; and lack of creative destruction of entrepreneurial innovation, 47, 52, 62–63, 85; Schumpeter's nightmare of stifling of innovation via, 41–42, 87–88; of scientific publishers, 132; solidification of, 46–47; as transitional phenomenon (Schumpeter), 52. *See also* antitrust actions; competition; "kill zone" (buyouts of start-ups); monopolies
markets: data markets as failure, 94–97, 103–104; exclusion right to

markets *(continued)*
data and, 95–96; "lemons" problem and, 95; scarce resources and, 95, 97; transaction costs and, 96–97. *See also* digital marketplace
Marshall Plan, 42
Massachusetts Miracle, 106
Mastercard, 68
Mechanical Turk, 151, 152
media: and closed vs. open government data, 138–139; failure to invest in the digital marketplace, 27; free press, principle of, 1–3; shift of power to Big Tech, 8
medicine: cancer research and treatments, 22, 27, 128; and reuse of data, 122; Silicon Valley startups causing difficulties for digital laggards in, 90. *See also* COVID-19 pandemic; drugs and drug companies; health care
Mejias, Ulises Ali, 153
Mercedes, 28
Merkel, Angela, 4
message services, 27, 59
meteorology, 122
Microsoft: abandonment of data market by, 96; acquisition of start-ups ("kill zone"), 60, 99; antitrust action against, 67, 83; Azure cloud platform, 32, 73, 96; Chinese regulation of, 113; COVID-19 pandemic and open data initiative of, 16, 146; GitHub, 33–34, 60, 116–117; as hyperscaler, 73; Internet Explorer, 67; market concentration of, 47; Office, 67; Windows, 67, 107–108. *See also* narrative of achievement
military: Georges Doriot and, 43; and GPS, 137–138; hosting the "DARPA Grand Challenge" for autonomous vehicles, 18–19; and Raytheon, 106; and telecommunications monopolist AT&T, 165
Minecraft, 60, 99
minimal data use. *See* data protection, privacy laws (minimal data use)
mobility data: and the COVID-19 pandemic, 3, 16, 146; regional, and COVID-19 pandemic, 3; smartphone apps for tracing infections, Big Tech blocking, 3–5
Moderna, 126
Mojang, 99
monopolies: AT&T as, 165–166; calls to address the issue of, 84–85; Chinese, 63–67, 76–77, 154–155; "creative," 62–63; and infrastructure services, 33; the "kill zone" and formation of, 60–61, 62; as slowing the pace of innovation, 76–77, 79, 85; Peter Thiel on formation of, 61, 62; as transitional (Schumpeter), 40–41, 52. *See also* antitrust actions; competition; data access mandates; market concentration
monopolist informational power. *See* informational power asymmetries
monopoly rents, 91, 97
Moore's Law, 31
Mozilla Foundation, 145
Musk, Elon, 34, 85, 165
MySpace, 46

N26, 27
Nadella, Satya, 8, 60
narrative of achievement (Big Tech): algorithms claimed to be issue in, 31, 33–34, 35; computing power claimed to be issue in, 31–33, 34–35; data colonialism and, 160–161; exclusive access to data as the truth behind, 30–31, 35–36; extraordinary human intelligence claimed to be issue in, 31, 34–35; as failing to discuss information asymmetries, 29–31, 35–36; *How Google Works*, 29–30; moral legitimization of power provided by, 35; representatives sent to public conferences, 35–36, 61–62; as scripted hoax, 30–31; Peter Thiel's characterization of as "propaganda," 61–62
NASA, 129
national champions, 64, 65–66, 69, 73–74, 76, 77, 161–162
National Economic Council, 84
natural disasters, 145
Nazi Germany, 43, 129
NEST thermostat, 121
Netflix, 27, 35, 47, 121
net neutrality, 8
Netscape, 46
NGOs (nongovernmental organizations): open data and, 131, 135–136, 144–145; using government open data, 142
Nigeria, 13, 163. *See also* data colonialism
Nintendo, 56
Nixon, Richard, Watergate, 138–139
Nobel Prizes, 44, 96, 167
Nokia, 46, 63
nonpersonal data: bureaucracies and failure to recognize, 140; consent as issue in, 101, 148; data protection laws blocking access to, 99–100, 148–149; depersonalization of data, 111–112, 120, 140; digital dividend of access to, 148; guidelines and technologies in development for, 120; smartphone apps for tracing COVID-19 infections, Big Tech blocking of, 3–5; as subject to mandated data access, 111–112, 120; uncertainty about status of, 101–102, 120. *See also* data access mandates; data protection laws; individuals exporting their personal data to digital platforms
Noveck, Beth, 129–130, 139–140, 142
Nuance, 60

Obama, Barack, 82, 140, 142
Ocado, 22
OECD (Organization for Economic Cooperation and Development), 136
Office (software), 67
OpenAI, 34
open data: overview, 129–131, 144, 147–149; access from anywhere, 147–148, 149; and data as public good, 127, 130; and data protection laws, 130, 148–149; data standards and, 129; democracy and good governance and, 129–130, 138–139, 147–148, 168; empowerment and,

open data *(continued)*
130, 149; enforcement of, 143, 144; fact checking and, 136; implementation as the issue in, 131; multiple uses for the same data, 133–134, 145; NGOs (nongovernmental organizations) and, 131, 135–136, 144–145; origins of idea, 129; the private sector and, 131, 142–144, 145–146; regulations and, 144; and the right to access information (UN International Covenant on Civil and Political Rights), 130–131; technical capacity vs. social reality and, 148. *See also* open data initiatives; open exchange of information; open government data; open science research

open data initiatives: Barcelona telecom (Telefonica), 15; CERN (European Organization for Nuclear Research), 134–135; the COVID-19 pandemic, 15–16, 126–127, 133, 136, 146; EU public transport timetables, 143; GPS, 137–138; Human Genome Project, 127–128; the World Wide Web as, 134–135

open exchange of information: as the driving force of the American Revolution, 1–3, 5, 8, 130, 168; Silicon Valley success as based on, 107–109, 125

open government data: overview, 131; the advent of the World Wide Web and, 129; chief information officers (CIOs) in charge of, 139–140, 142; confidentiality principle as obstacle to, 138, 140–141; data literacy training and, 141; data protection laws constraining federal authorities, 139; and democracies, health of, 129–130, 138–139; economic value of, 138; the economy and innovation stimulated by, 143; ecosystem of information and communication needed for, 141–142; enforcement of, 143, 144; freedom of information rules as adversarial vs. integral duty, 139; GPS, 137–138; mindset required for, 141, 144; the Obama administration and, 140; private-sector mandatory data access as necessary adjunct to, 142–144; size of governmental unit and difficulty with, 140–141; structural impediments to, 139–141; US online platform for (data.gov), 142

open science research: overview, 131; and the collaborative project of science, 134; and COVID-19 pandemic, 16, 126–127, 133, 136; and data as public good, 127; data standards and, 134, 136; data underlying, importance of open access to, 133; decentralized infrastructures for, 136–137; digital distribution and, 132; FAIR (findable, accessible, interoperable, and reusable), 133; government as facilitating, 127–128; government funding requirement for, 132, 133; multiple uses for the same data and, 133–134; origins of the idea of open data and, 129;

private scientific publishers and lack of, 131–132, 133; reproducibility and verifiability and, 133–134; resistance to, 133; reuse of data and, 122; science journals turning to, 132–133; scope of available data, 135; technical capabilities for sharing, 134
open-source algorithm libraries, 33–34
OpenStreetMap, 144–145
"Osaka Track" international data cooperation, 163

Page, Larry, and centralization of power, 8
Pakistan: data colonialism in, 13, 153; and mandated data access, 163. *See also* data colonialism
Palantir, 61
Pariser, Eli, 8
patents: banned on human gene sequences, 127–128; of Bell Labs, mandated access to, 166–168; open government data and, 140; shrinking number of companies registering majority of, 51; Silicon Valley and sharing of, 108
Pax Americana, 154
PayPal, 27, 35, 61, 78
Pennsylvania Gazette (Ben Franklin's newspaper), 1–2, 8
personal computers (PCs), betrayal of promise of empowerment through, 7–8
personal data. *See* individuals exporting their personal data to digital platforms; nonpersonal data
Peru, 153
pharmaceuticals. *See* drugs and drug companies
Philippines, 153
Philippon, Thomas, 51–52
Phoenix, Arizona, 21
Pichai, Sundar, 4–5
platform hermeneutics, 25–28
PlayStation, 56
"polar inertia," 50
Polaroid, 105–106
political atmosphere: and betrayal of promise for change, 7–8; polarization and radicalization as fueled by Big Tech, 148, 164. *See also* data access mandates; data colonialism; governments
Pong, 56
Pony.ai, 20
post-acquisition lockup measures, 83–84, 86, 87
Postal Service Act, 3
PriceGrabber, 26
privacy laws. *See* data protection, privacy laws (minimal data use)
private sector: open data and, 131, 142–144, 145–146; open government data requiring mandated data access from, 142–143; using open government data, 142. *See also* Big Tech; digital laggards; small and medium-sized companies
Public Data Explorer (Google), 136
public health. *See* COVID-19 pandemic
public offerings of companies (IPOs), 58, 60, 83
public transport, 143, 147–148

Quantum, 108

R (software), 33
Raytheon, 105–106
Reagan, Ronald, 137
real estate, and open data, 145
regulation of Big Tech: Australian law requiring licensing for local news content, 113; Big Tech as fine with current attempts at, 11; EU rules preventing buyout of European start-ups, 87; Facebook fines and external audit (2019), 80; globalization and lack of, 154; growing calls for, 8–9, 15, 80–86; information asymmetries growing despite attempts at, 11; national laws, Big Tech complying with, 113–114; profitability of "kill zone" acquisitions as political poison for, 86; proposals for incremental regulations to foster alternatives to "kill zone" acquisitions, 83–84, 86, 87; US legislative gridlock and difficulty of, 85–86; as utilities, 84–85. *See also* antitrust actions; data access mandates; data protection laws; regulations
regulations: freedom of information rules, 139; national minimum wage and working condition laws not enforced for clickworkers, 152; open data, 129
remote sensing data, 129
renewable energy, 147
research. *See* open science research
retailers: price comparison services, 26, 47; recommendation algorithms, 22; robots serving customers, 63–64; Silicon Valley startups causing difficulties for digital laggards in, 90
reuse of data, 121–123
Robinhood (fintech company), 27
robo-taxi services, 21
robots, 50, 63–64
Rosenberg, Jonathan, 29
Route 128 (Boston), 105–107, 108, 109, 125
Russia: and data access mandates, 11–12; Gazprom, 68. *See also* Soviet Union

Samsung, 63
Sanders Associates, 55–56
Santa Clara Valley, 105, 106. *See also* Silicon Valley
satellites, 129, 137, 145
Saxenian, AnnaLee, 106, 108, 109, 117
scaling, 73, 78
Scandinavia, open public records in, 99, 138
Schmidt, Eric, 29, 61–62
Schumpeter, Joseph Alois: on anti-capitalist intellectuals, 40; biographical information, 37–38, 40, 41, 45, 50, 53–54; fear of socialism of, 41–42, 46; on "founder's profit," 45; on the importance of entrepreneurs, 39, 78; on Keynesian economists, 40; as microeconomist, 39–40; on monopolies as transitional, 40–41, 52; on the need for capital to finance innovation, 41, 85; nightmare prognosis of the decline of capitalism due to

deprivation of resources needed by entrepreneurs, 41–42, 46, 79, 87–88, 97, 123. *See also* creative destruction of entrepreneurial innovation (Schumpeter); Schumpeterian competition; Schumpeterian innovation
Schumpeterian competition: China and, 65; coopted by structure of venture capital market, 60
Schumpeterian innovation: destroyed by the kill zone, 60–61; and mandated data access, 103
science research. *See* open science research
scientific publishers, private, 131–132, 133
Seagate, 108
search engines, 22
Sega, 56
seismology, 122
SenseTime, 35
Shapiro, Carl, 6
Shazam, 60
Shockley, William, 44, 167
short-term rentals, 47
Siemens, 112
Silicon Valley: coining of term, 106; as contributor to US presidential campaigns, 81; and culture of collaborative innovation, 107–109; decentralization of company structures in, 107, 108; establishment of, 105, 106; and open-source algorithms, 34; opportunity for other regions around the world to become like, 109, 125; William Shockley as jump-starting, 44, 167; workplace culture of, 29; workplace culture of, Big Tech breaking with, 107–109, 125. *See also* Big Tech; "kill zone" (buyouts of start-ups)
Singapore, 4–5
Skype, 47, 48, 99
small and medium-sized companies, and mandated data access: benefits of, 112; follow-on innovation after, 167; measures to assist with, 119–120; percentage of data required to be shared by, vs. large companies, 112. *See also* digital laggards
smartphone app stores: as business model, 25; as ecosystem, 141–142; and exclusive use of data, 30; market concentration in, 46–47; US attempt to ban Chinese apps from, 66
smartphones: COVID-19 tracing apps crippled by Google/Apple, 3–5; GPS, 137–138; Huawei, 73. *See also* individuals exporting their personal data to digital platforms
Smith, Adam, 39
Snowden, Edward, 70
social democrats, 46
socialism, Schumpeter and fear of, 41–42, 46
social media: betrayal of promise of empowerment through, 7–8; clickworkers and content screening on, 151; current trivial usage of, 148; informational power asymmetries and, 30; and "kill zone" acquisitions of start-ups, 59; national laws regulating, 113–114; open data and

social media *(contined)*
beneficial use of, 148; political polarization and radicalization fueled by, 148, 164; and superiority of Chinese products, 155. *See also* Facebook; WeChat
social sciences, open data and, 135
Softbank, 48
Sony, 56
South Africa, rejection of the "Osaka Track" data cooperation proposal, 163
South America: clickworkers in, 153; data colonialism in, 159, 161–162; open science research in, 133; private sector donations of open data, 145
Soviet Union, 137. *See also* Cold War; Russia
speech recognition software: acquisition of start-ups, 60; machine learning and, 23; training of, 27, 145
Spotify, 30, 35, 59–60, 90, 99
Standard Oil, 81
standards: European GAIA-X cloud project and unification of, 73; proprietary, as maintaining informational power asymmetries, 120; and "Tech Cold War" (China and the US), 158, 159. *See also* standards, open
standards, open: and collaborative innovation of Silicon Valley, 107; and open science research and data, 134, 136; and operation of mandated data access, 118, 119–120; and origins of open data (in geophysics), 129
Stanford University, 105, 107
start-ups: Chinese, 63–66; data protection laws as hampering, 11; decreasing as percentage of all companies, 51; "fake it 'til you make it," 52; minimal viable product, 57; special role of the investor in, 41; traditional "biological exit strategy" and, 57; "unicorns," hope for shaking up the status quo with, 89–90. *See also* "kill zone" (buyouts of start-ups)
steam engines, 124
Stockholm, 99
storytelling. *See* narrative of achievement (Big Tech)
streaming services, 27, 35, 47, 121; Chinese, 63
Streetspotr, 151
Street View (Google), 27–28
Stripe, 27
Sun Microsystems, 45
Super Mario, 56
Suri, Siddharth, 150–151, 152
surveillance: China and government data access for, 65; EU and fear of surveillance state, 69–70, 72; surveillance capitalism, 8, 70
sustainability, open data and, 147
Sweden: and data access mandates, 98–99, 138; Spotify, 30, 35, 59–60, 90, 99
Switzerland, CERN, 134–135

talent: access to data as issue for, 53, 98; and acquisition of start-ups, 58–59; Big Tech's ease of hiring, 53, 97–98; and Big Tech's narrative

of achievement, 31, 34–35; and data colonialism, 162; diversity and, 77; global brain drain, 97–98; noncompete clauses, 107; and the open exchange of information of Silicon Valley, 107; opportunity as attracting, 98; and the "Tech Cold War" (China and the US), 156
Taylor, Frederick Winslow, 151
TCP/IP protocol, 50
"Tech Cold War" (China and the US): overview, 66; Biden administration and, 156; China's economic and infrastructure aid to client states, 155, 158; China's projection of power and, 66, 158; China's rise to superpower status, 154–155; China's structural advantage in, 156; and data access mandates, 11–12; and data colonialism, 159; decoupling of networks, 155; decoupling of supply chains, 156–157, 158; definition of, 154; global division into two camps, 155, 157–159; hegemony as goal in, 158–159; and the "Osaka Track," rejection of, 163; Pax Americana of technology and, 154, 155; skeptics of the metaphor, 157; skirmishes between surrogate forces, 155; standard setting and, 158, 159; Trump administration and, 155–156; US protectionism and, 66, 158; and war, potential for, 157. *See also* data colonialism
"techlash," 9
technology: governments supporting specific, vs. supporting capacity for innovation, 64–65, 74–79; and informational power asymmetries, 25; of mandated data access, not to be specified by law, 119; for nonpersonal data, 120; for open data, vs. social reality, 148; for open science research sharing, 134; Pax Americana of technology, 154, 155. *See also* information technologies (IT); "Tech Cold War" (China and the US); *specific technologies*
telecommunications companies, 27, 51, 84–85. *See also* AT&T
Telefonica (Barcelona), 15
temperature data, mandated access and, 119
Tencent: and centralization of power, 8; emergence of, 63; government access to data of, 65; WeChat, 27, 30, 63, 155–157. *See also* narrative of achievement
Tenpay, 66
Tesla, 20
Thiel, Peter, 61–62, 85
"Thucydides trap," 157
TikTok, 155–157
Tinder, 50
Toffler, Alvin, 49
tourism, 90
Toyota, 28
trade secrets, not subject to access mandates, 111
tragedy of the commons, data not subject to, 14
transistor, invention of, 44, 166
transparent public sector. *See* open government data
transportation: logistics, 122, 137–138; public, 143, 147–148

travel platforms, 26, 47, 109–111
Trivago, 90
Trump administration, 9, 61, 66, 82, 155–156
Turing, Alan, 7
Turkey, regulation of YouTube, 113
23andMe, 26
Twitter, 59, 82

Uber, 35, 150–151
Uber Eats, 26
United Kingdom: British colonialism in the US, 1–2, 168; and data access mandates, 110; and open science research, 133
United Nations: Conference on Trade and Development (2019), 162–163; International Covenant on Civil and Political Rights, 130–131
universities: talent leaving for Big Tech, 97–98; tuition increases in, 51. *See also* open science research
Unix operating system, 167
US Congress: banning patents on human gene sequences, 128; and data protection laws, 139; and Facebook, 80; freedom of information laws, 139; regulation of Big Tech as unlikely due to gridlock of, 85–86
US Department of Defense, 18–19
US Department of Justice, and mandated data access, 109–111, 118, 123
US Federal Trade Commission (FTC): critics of Big Tech appointed to, 9, 84–85; fine imposed on Facebook, 80
US National Institute of Health, 126, 127, 128
US Post Office, 2–3
utilities: closed data pools for, 117; regulation of Big Tech as, 84–85

Varian, Hal, 6–7, 109
Venezuela, clickworkers in, 153
venture capital: coining of term, 42; as enabling high capacity of innovation, 45–46, 49, 85; establishment of, 45; as failing in reproduction and disruptions of large-platform successes, 48; as innovation, 48–49; and the "kill zone" (buyouts of start-ups), 48, 58, 60, 86; and the open exchange of information in Silicon Valley, 107; postwar era and easy access to, 45; and public offerings of companies (IPOs), 58; and risk of radically new business models, 28; and Route 128 (Boston), 105; as thwarting innovation, 60, 85; "venture debt" bank loans as alternative to, 83
venture debt, 83
Vestager, Margrethe, 9, 67–69, 86
video game consoles, 55–56
Vimeo, 47
Virilio, Paul, 50
voice recognition software. *See* speech recognition software
von der Leyen, Ursula, 68–69, 89, 117

Wang, 105–106
Warren, Elizabeth, 80–81, 83
Watergate scandal, 138–139

Waymo, 19, 20–21, 22, 151
Waze map service, 59, 81, 85
Weber, Max, 6, 8
WeChat, 27, 30, 63, 155–157
WhatsApp, 47, 48, 59, 81, 82
Whole Foods, 81
Wii, 56
Wikipedia, 144
Windows, 67, 107–108
Wise, 27, 90
Wooldridge, Adrian, 9
World Bank, 136
World Values Survey, 135
World War II, 41–42, 43, 44, 55
World Wide Web: as decentralized network, 135; establishment of, 50, 134; the Internet transformed by, 134; open government data and, 129
Worms (bank), 43
Wuhan, China, 126
Wunderlist, 48
Wu, Tim, 8, 9, 81, 84, 85

Xbox, 56
Xerox PARC, 107–108
Xiaomi, 154–155

Yahoo, 113
Yitu Technology, 35
YouTube, 47, 48, 113
YY, 63

Zalando, 90
Zoom conferences, 50
Zoox, 20
Zuboff, Shoshana, 8, 70
Zuckerberg, Mark, 8, 59, 80, 81